Ready, Set, Done.

How to Innovate When Faster is the New Fast

by

Jim Carroll

OBLIO Press, www.obliopress.com

Carroll, Jim, 1959-
     Ready, set, done : how to innovate when faster is the new fast / Jim Carroll.

ISBN 978-0-9736554-2-1
     1. Technological innovations-Management.   2. Creative ability in business.   I. Title.

HD45.C38 2007   658.5'14   C2007-905266-5

Production Credits:
Editors: Robert Mustard, Mark Jeftovic, Christa Carroll
Cover Design: Phil Emery / Focused Creative Communications
Photography: R. Kelly Clipperton

Printed in Canada

Table of Contents

CLOSING

To Christa,

Willie and Thomas,

for putting up with fast

when it's fast,

and helping to slow things down

when they need to be slow.

Ready, Set, Done. • How to Innovate When Faster is the New Fast

Fast

Sometimes when you innovate,

you make mistakes.

It is best to admit them quickly,

and get on with improving your other innovations.

Steve Jobs

Faster

If everything seems under control,

you're not going fast enough

Mario Andretti

Ready, Set, Done. • How to Innovate When Faster is the New Fast

Plasma People, Cardboard People

Why is faster the new fast?

PICTURE THIS: it's a large meeting room, in a big, non-descript city hotel. It's early evening, cocktail hour for the folks arriving at a national conference.

There's a lot of laughter, smiling, hand-shaking, as old acquaintances are renewed. It's an annual event for an industry that goes back many, many years. They are involved in designing specialized promotional displays for use in the retail sector.

Upon entering the room, you notice something – well, odd.

There's a group of folks on one side, you guess by their age, primarily baby boomers. Then there's a big, empty space down the middle of the room, and over on the other side there are a bunch of young people, chatting quietly. There just doesn't seem to be a lot of mingling between the two groups.

Intrigued, you endeavor to find out why there is this 'great divide', and walk over to chat with the baby-boomer crowd.

It turns out that they are the vanguard of this particular industry. For years, they've used tremendous amounts of creativity to come up with very unique, in-store displays – those cardboard display cases that you often see near the cash register. They've helped consumer goods companies and retailers move a lot of product. They're busy sharing their stories as to what they've been doing for the last year. They're also here to learn about the latest leadership ideas, innovation techniques, and other insight that can help them keep their "edge" in the display market place.

You decide to move on to find out who the folks on the other side of the room are. It turns out that this younger group is having a similar conversation about in-store displays but their view of the promotional displays of the future is totally different. They foresee a highly interactive shopping experience where consumers enter a store and utilize a wide variety of very cool, very sophisticated technology.

Their discussion is about picking up a product in aisle 3, and having an RFID tag that is embedded within the product link to the plasma TV at the end of the aisle. This causes a quick promo-hit to run, letting the consumer know about a related, discounted accessory over in aisle 27. The consumer then picks up an item in aisle 12, which causes his Blackberry to buzz minutes later reminding him that the last time he purchased that product, he forgot to pick up the batteries for it. As the consumer goes through the checkout, they use the same Blackberry as a personal bank-card, simply swiping it over a wand, keying in their PIN, and then they're done.

This younger group has a lot of really cool ideas, yet they express to you their frustration with some of the veterans in the industry that they just don't seem to be interested in what they are talking about.

Fascinated, you begin to think a bit more about the dynamics in the room. You go back and talk with the folks on the other side of the room – let's call them the "cardboard people" – and ask them about this new world of interactive in-store plasma TV's, and the "plasma people."

They dismiss the concept with a wave of their hand, even though the technology that the 'plasma people' refer to already exists today.

This younger group has a lot of really cool ideas, yet they express to you their frustration with some of the veterans in the industry that they just don't seem to be interested in what they are talking about.

In an instant, I realize that there are many organizations in the world today that are in the same place as this organization. They've got their cardboard people, and their plasma people, and the two just don't see eye-to-eye.

If only you could get both groups to work together, combine their creative experience and insight – that would be magical.

A senior executive in the room of this conference was over-head saying, *"… people are engaged in survival tactics because they don't know what to do next …"* There's a lot of this type of uncertainty within many industries.

The rate of change today – whether with business models, product lifecycles, skills and knowledge, marketing method-ologies or customer support concepts – is speeding up. We live in a world where being faster is better than being fast.

That's why innovation is the most important word that you need to be thinking about. Innovation is all about adapting to the future – and if the future is coming at you faster, then you need to innovate faster. Innovation shouldn't be about trying to survive the future – it should be about thriving.

There are certain undeniable truths about the future:

- **It's incredibly fast.** Examples surround us. Product life-cycles are collapsing. Half of what students learn in their freshman year about science and technology is obsolete or revised by their senior year. There are furious rates of new scientific discovery. Time is being compressed.

- **It involves a huge adaptability gap.** The current generation of management within many organizations – baby boomers – have participated in countless "change management workshops," reflecting the reality that many of them have long struggled with change. Gen-Connect – today's 15 and under – will never think of change management issues. They just change.

- **It has a huge instantaneity.** The next generation thinks at video game speed. They scan 12 feet of shelf space per second and to them most news becomes old news within 36 hours of emerging. Everything is becoming instantaneous.

- **It hits you most when you don't expect it.** Old-hat ideas are always around. These ideas are exchanged and often percolate for a while, only to be dismissed as unrealistic. But add in the rapid emergence of new technologies, and suddenly these dismissed old-hat ideas get adopted. The next thing you know, the concept has steamrollered you into submission.

- **It's being defined by renegades.** Increasingly, the future of many industries are being defined by industry expatriates. When a real innovator can't innovate within a company, they step outside, form a startup, and spark massive industry change on their own. Before you know, they've reinvented you.

- **It involves partnership.** Old business models involved asking, "what can we do to run our business better?" The

When a **real innovator can't** innovate within a company, they step outside, form a startup, and spark massive industry change on their own.

new business model involves asking: "What can we do to run our customers, suppliers and partners business better?"

- **It involves intensity.** We must learn to run our business at video-game intensity: in fast paced markets, we need fast paced business capabilities.

- **It's bigger than you think.** Think GoogleCar or iPlane. Complacency is a dangerous thing, particular when every organization is faced with constant, relentless external innovation from unexpected competitors.

- **It involves everyone.** With rapid change, everyone in an organization must innovate. To adapt to the future, you can't rely on "brilliant ideas" from a bunch of "innovation elitists," or think that only special people can "do" innovation. Thriving in the future has a leadership that involves everyone in innovation. No idea is too dumb, no opportunity is too small.

- **It comes from experiential capital.** With a fast future, you've got to learn and relearn. Corporate equity isn't just money: it's the cumulative experience and knowledge of the team.

- **It requires a combination of skills.** It involves figuring out how to get the plasma people and the cardboard people to use their combined strengths, insights and capabilities to do something magical.

- **It involves understanding how to link rapidly emerging trends to radical ideas in order to do things differently.** And it involves doing that fast – actually, faster than fast.

Forget about the concept of innovation as simply involving the design of cool new products.

In the high-velocity economy, where faster is the new fast, it's your ability to adapt, change, and evolve, through a constant flood of new ideas, that will define your potential for success.

That's what innovation is, and that's the focus of this book.

Stuck in a Rut?

What's your tin can?

MAYBE THE first step in rethinking innovation, is to think about what's been happening with tin-cans. Have you been to your local grocery store as of late? Did you see the new StarKist Tuna plastic re-sealable pouch?

That little package – a new product innovation if there ever was one – is responsible for almost $200 million in new revenue since it first hit the shelves.

That's not displaced revenue, but entirely new revenue that didn't exist before.

It's a big change – and it took a long time to come about. After all, StarKist sold tuna for 110 years in the same old way – in a tin can. Yet they finally managed to come up with something new, and the results are stunning.

The new tuna pouch is a good segue into what is perhaps one of the most important issues that you need to deal with in order to lead innovation – getting people within your organization out of their tin-can rut!

How many are still stuck in a 110-year old rut? Still delivering a tin-can day in, day out, with no desire to change? Still focused on cardboard, or some other concept or device or method or idea that is clearly linked to the past. Unfortunately there are quite a few – who wants to try anything new today?

Ask yourself this:
are people in your
organization stuck in a rut?

It's very easy for people to lose their drive, their courage to go forward, and their willingness to change. They're still making tin cans, when new re-sealable pouches could revolutionize who they are and what they do!

That's why the StarKist story is so important. Here's an organization that has somehow shaken away the complacency that enveloped it for over a century. It has woken up to the opportunity that comes from real innovation. It is all part of a reawakening that is underway throughout the world, in which people are realizing that there are tremendous opportunities that come from doing new things.

One example is establishing new business partnerships that help you innovate with product design. Over the last few years, food and packaging companies have come together in a partnership that redefines how new products are developed. Packaging companies, previously restricted to the sidelines, now take a lead role in the development of new products. Food companies, who used to be the only ones responsible for new products now realize, that if they are willing to open up their minds to a new way of doing things, they can see some darned powerful results.

Yet it took a long time for such innovative partnerships to take root.

It took Star-Kist 110 years!

Ask yourself this – are people in your organization stuck in a rut? Is your organization still making "tin cans"? Are they locked in routine, with the same stale ideas, old methods of operating, concepts and routine that lead to lousy customer service, inefficient business processes, and all kinds of other olden-days ways of doing things?

If so, what should you be doing instead?

What are your tin cans, and can you get rid of them?

It's an important question to be addressed in a world that continues to evolve at an ever-increasing pace.

I've long been using the tin-can story when speaking to organizations about the challenges of change. That's because the simple concept of being stuck with a "tin-can" provides people with a simple, realistic metaphor that helps frame their attitude towards change.

One of my most memorable experiences involved a presentation to a large group of military personnel; my keynote focused on the rapidly changing technologies, methodologies, concepts and structural, organizational change that is impacting every military organization worldwide.

I debated prior to my presentation: should I use the "tin-can" story?

I did – and was pleasantly surprised to learn that the metaphor became a common theme throughout the rest of the three day meeting. From what I heard, senior military personnel, many of them in uniform, prefaced their own presentations to the audience with the need to think about, confront, and examine their tin-cans, because that would be one key method by which they could ensure they could adapt to the rapidly changing world of the 21st century!

Assassination Attitudes

Are you at risk from the innovation killers?

"The definition of insanity is doing the same thing over and over again and expecting a different result each time."

– Albert Einstein

Are you stuck in an innovation rut? Lots of tin cans?

Maybe so: many organizations have become so focused on getting things done that they've lost their ability for any sort of creativity. That's why Einstein's comment is so appropriate; in a world of rapid and constant change, many people still manage to think that they can get away with routine.

That's why you should undertake an "attitude inventory" during your next staff, board or executive meeting. It's easy enough to do. At your meeting listen for these innovation-killer phrases. Every time you hear one, score one point (except for special scoring where noted!):

- **"We've always done it this way."** Chances are that you will hear these words early on in any meeting. This is the worst possible phrase that can be used by any organization; so bad, in fact, that you score extra points for this one. It means that you've got some pretty thick organizational sclerosis that is clogging up your ability to try to do anything new. If your team members are in this frame of mind, it's time for some pretty dramatic action! *(Score one point – unless phrase is used within first five minutes of meeting then score a bonus of 5 points!")*

- **"It won't work."** In this case, pessimism has become the cornerstone for success. Any new idea or initiative is immediately shot down and relegated to the failure heap even before it's had a chance to be looked at. That's not a good way to take yourself into the future, and hence, such cynicism should be outlawed. (*Score one point*)

- **"That's the dumbest thing I ever heard."** Anyone who dares to be different and suggests anything new and innovative will have the wind taken right out of their sails as soon as they hear this phrase. Best to address this issue by ensuring that people understand the cardinal rule of innovation: it's better to have 99 dumb ideas and one great idea, than to not have any ideas at all. (*Score one point*)

- **"That's not my problem."** There's a bit of a teamwork issue here, don't you think? In a world of massive complexity and ever-increasing challenges, an organization needs to have a culture in which everyone is able and willing to pitch in, regardless of what might be involved and what it might take. (*Score one point*)

- **"You can't do that."** Often this happens when someone comes up against ingrained organizational behavior, or smacks their head into the rules and regulations defining day to day operations. Most often we encounter this reaction when governance issues rise to the forefront. Structure is important, but it shouldn't stand in the way of innovation. (*Score one point*)

- **"I don't know how."** At a time when careers are changing overnight, this might indicate that you are doing rather poorly in the areas of education and professional development. In today's age, no one should be in a position in which they lack necessary skills. The only way to ensure that your organization can meet the challenges ahead is if everyone is in a mindset in which they are prepared to take on anything, and are willing to do what is necessary to learn what's needed to get things done. (*Score one point*)

"You can't do that." Often this happens when someone comes up against ingrained organizational behavior, or smacks their head into the rules and regulations defining day to day operations.

- **"I don't think I can."** Self-confidence can be lacking in many people today, given that they are finding so many new demands on their skills. If you encounter this phrase, it is probably best to nurture a culture that supports risk; people should know that they can take on something new, and not suffer from adverse consequences if it doesn't quite work out. (*Score one point*)

- **"I didn't know that!"** Lack of proper communication is behind the failure of many new initiatives. When someone uses this phrase, it indicates that you've got a pretty serious problem on this front. Make sure that you've got a culture that is collaborative and open; in which information sharing is encouraged and information-hoarding is punished. (*Score one point*)

- **"The boss won't go for it."** Innovation starts at the top, and the leadership must set the proper tone in order to make it possible. If this phrase is used, you've really got some serious trouble! (*Bonus phrase – 5 points!*)

- **"Why should I care?"** Several recent surveys have shown that anywhere from 50% to 75% of staff in a typical organization will jump ship if given the chance. Another

You might have some pretty thick organizational sclerosis clogging up your ability to try to do anything new.

survey showed that at least 40% of people are already thinking about a better job – on their first day in a new job! If you encounter this phrase, it means that people aren't engaged; they don't share your sense of passion and purpose. You've got to do more to build a collective spirit and an overall culture that supports everyone moving forward in the same direction. *(Extra bonus points: 10)*

While there is no magic scientific formula, it's probably a good guess that if you score over 10 points in one meeting, you've got a big problem. Score over 20, and you are innovation-dead. Over 30 points, and you might as well shut the whole game down, close up shop, and lock the door; you don't deserve to be in operation.

Through the years, the material for this chapter has become one of the most viewed pages on my Web site. Whether I'm on stage in front of a few thousand people or in a small management get-together, I can see people nodding their heads up and down when I talk about the innovation killers.

Clearly, the phrases are in common use, and strike a chord with everyone. The fact that so many people recognize them truly shows that many organizations have some pretty big innovation challenges.

Velocity

Velocity

The real reason
that innovation is critical

YOU KNOW that innovation is important, but do you
really know why? It's because of "high-velocity change"
– quite simply, a world in which science, ideas, business models, careers, and just about everything changes at a
furious pace. Velocity causes significant change, and to keep
up with significant change, you need to be innovative: that is,
you need to do things differently on a fairly continuous basis.

In this section, we'll take a look at the concept of velocity
from a variety of perspectives:

- Velocity is driven by furious rates of scientific advance.
  In "**Nano Numbers,**" we outline that as science evolves
  quickly, so does the discovery of new ideas that result in
  innovative new products.

- Velocity is also increasing because of the emergence of
  the infinite idea loop, which not only drives the more
  rapid emergence of new science, but also provides for the
  rapid sharing of ideas. It's by "**Tuning Into the Future**"
  and learning to take advantage of the ideas within the
  idea loop that you can begin to discover innovative ideas.

- In "**It's So Yesterday!**", we suggest you take a look around
  you and do the "10 Things Test" to get a sense of how
  velocity is changing our world in small, subtle, yet significant ways.

If you don't keep up with high-velocity change, you'll find an increasingly negative impact on the value of your brands or your image.

- To keep up with velocity, you'll have to think about whether your business or organization might need to run at the same speed as found within the video game industry. In "**Game Time!**", we outline why the key word to think about is "intensity."

- It's easy to think that you won't be impacted by velocity: after all, every industry is different, and some evolve quicker than others. Yet, in "**Speed Freaks,**" you'll see how a sausage maker, a heart valve company, and a board game manufacturer are all being impacted by high-velocity change.

- If you don't keep up with high-velocity change, you'll find an increasingly negative impact on the value of your brands or your image. In "**Hero to Zero,**" we'll challenge you with the question as to whether your brand could become something from the olden days.

- All of which leads to a very simple question: are you prepared to deal with velocity? In "**Time Travel,**" we'll challenge you to ask yourself whether you really understand the rate of velocity that is impacting you.

Nano Numbers

Science driven innovation- staggering numbers, massive opportunities...

I F YOU want to understand the concept of high-velocity change, you might do well to look at an iPod. For it's within the rapidly changing technology of the iPod product line that we can witness first hand how rapid rates of scientific change drive furious rates of product innovation, which drives ever faster business change.

Certainly iPod's are one of the hottest consumer technologies in the marketplace today. But it's more than just a cool piece of electronic hardware that plays music. The product line provides ample evidence that we live and work in a world in which massive, sudden, wrenching change will become the norm, not the exception. Here's a brand that, within only five years, has seen several generations of product, each one making the prior version obsolete because of the rapid advancement of science.

When Steve Jobs introduced the new iPod Nano, he announced that the company would no longer manufacture the larger sized iPod Mini. Yet up to that very moment, the iPod Mini had been one of the coolest MP3 players around, generating annualized revenue of over $1 billion US. The iPod Mini itself was a relatively new product, having been introduced just one and a half years prior to the introduction of the iPod Nano! However, the iPod Mini met its instant demise with the introduction of the iPod Nano.

Both the original iPod Mini and the iPod Nano resulted from new scientific discoveries; in the case of the iPod Mini, it was the discovery of a single new chemical substance that permitted the miniaturization of the hard disk to fit within a small form factor. In the case of the iPod Nano, it was the rapid evolution of the science behind flash memory: quite simply, companies were quickly figuring out how to cram more and more memory into a unique technological device that had no moving parts.

That provides a good example of how the fundamentals of scientific advance can lead to rapid rates of product innovation. Every single aspect of the computer and technology industries are being impacted by rapid scientific change. Yet, countless other industries are similarly impacted. Consider, for example, the pharmaceutical industry, which is seeing the constant development of new drugs as a result of scientific R&D. Likewise, the agricultural industry is constantly changing as a result of new farming methodologies related to new pesticides and bio-crops. Then there is the energy industry. With everyone focused on renewable energy supplies, there are furious rates of new scientific discoveries occurring with wind, solar and other forms of alternative energy.

Science is at the root of all innovation in many industries: it drives new product discoveries, results in massive market shifts, and obsoletes existing products.

Just how quickly is science changing? Consider some of the mind-boggling statistics that lie behind but one branch of science: the world of chemistry.

An article by Joachim Schummer *"Coping with the Growth of Chemical Knowledge"* published in Educación Química in 1999, outlined the rapid rate of discovery occurring in the world of chemical sciences, specifically noting that:

- "… to be up-to-date in all areas of chemistry you would currently have to read about 2,000 new publications every day

One single new chemical substance led to the invention of the iPod Mini, which became a billion dollar market overnight.

- if you prefer to screen only the short abstracts, you must read 200 pages per day or about 70,000 pages per year

- furthermore, since the number of chemistry publications increases also exponentially, you need to double your reading capacity within the next 15 years …

- you must read 20 publications every day in order to grasp only 1% of the overall chemical publications!"

It is not only the impact on Chemical publications but also on the ever spiraling increase in the discovery of new chemical substances. Schummer went on to note that:

- "… the number of known substances has been growing exponentially since 1800, from some hundreds then to about 19 million today

- … the number constantly doubles every 13 years

- … we will have nearly 80 million substances in 2025, and about 300 million by 2050…"

That means in just a little over a decade, we will witness a 15-fold increase in the number of known chemical substances. The math becomes even more interesting: at this rate of growth, it is estimated that by the year 2100, there will be 5 billion known chemical substances, instead of the 19 million that exist today!

The numbers and the growth it implies are absolutely staggering. What's more so is when you realize that the discovery of a single new chemical substance can lead to the emergence of a billion dollar industry literally overnight – as it did with the iPod Mini.

That's why innovation is critical: given rapidly changing science, there is no doubt that we will continue to witness extremely high velocity with product innovation.

I've learned that I can walk into any industry and quickly spot the signs of science-driven high-velocity change. What do I look for? Rapid product development and changes to existing product lineups; new manufacturing methodologies; or the emergence of breakthrough products causing significant market upset.

To be innovative, you need to develop a capability to spot the scientific driven trends that might impact your career, company or the industry you work within. You can do this through direct observation and study, but sometimes, you'll discover a significant trend simply through a small, everyday event.

Right now, I'm witnessing a big change as a lot of "devices" or "things" take on a degree of connectivity. An example is bio-connectivity: we are seeing the emergence of a new category of remote, connected medical devices which permit the monitoring of patients from afar. This is a trend that has huge implications for the global health care industry, because home-health care is set to take on a major role as baby-boomers enter their twilight years.

In other cases, I'll spot signs of a significant future change, simply by listening carefully to the observations of others. It's a well-honed skill, and can be a useful one to develop.

For example, while driving my Jeep one day many years ago – the last vehicle I drove which featured a standard transmission – my youngest son quietly asked me from the backseat as to why I was using the "joystick."

Thinking it through, I came to the conclusion that the manner by which this next generation interacts with the world is very, very different – to such an extent, that we'll probably witness the disappearance of the steering wheel in our lifetime. It will be replaced by a new, highly interactive joystick, which itself is linked into highly intelligent smart-roads, in-vehicle distance and motion sensors, and other smart technologies that will make the driving experience of tomorrow completely unlike that of today.

Ready, Set, Done. • How to Innovate When Faster is the New Fast

Tuning Into the Future

The infinite idea loop

IF YOU want to understand the second reason why innovation is happening faster, it's because ideas are happening faster. And ideas are happening faster because of the emergence of what I've come to call the "infinite idea loop."

The impact? Everything around you, from basic science, to knowledge, career skills, business models, competitive challenges, the emergence of new markets – and just about everything else – continues to evolve faster than ever before. The rate at which change is occurring will only continue to speed up – because the velocity of change itself is increasing.

Why is this so? One primary reason is because innovation has moved from the corporate to the collective, a trend that is causing absolutely furious rates of discovery of everything that is new.

Think about it in the context of the discovery of new scientific knowledge. Fifteen years ago, the exchange of new ideas related to research and scientific advance in the world of medicine occurred at a rather leisurely pace, through conferences, journals and publications. Ideas took time; research occurred at a slow and methodical pace.

Today, new discoveries and knowledge are shared almost instantly, through online Web sites, discussion groups, blogs, e-mail and other instantaneous knowledge exchange tools. With this information exchange comes greater degrees of collaboration, sharing of insight, and research results. Ideas bounce off each other in an almost instant fashion – the globe has become one massive idea generation machine, with new ideas being created instantly, everywhere, on a continuous non-stop basis.

The result is an explosive growth in the discovery of new knowledge. Call it "Carroll's Constant," it's a measure of how quickly knowledge in any particular profession, industry or area of human endeavor is doubling. There are different rates in different fields, but suffice it to say it's a number that is constantly decreasing. For example in the medical field, some estimates suggest that the volume of medical knowledge is now doubling every eight years.

That's the type of thing that is happening everywhere. Today, we find ourselves in the midst of a global infinite idea loop, in which new ideas, inventions and innovations are shared faster than ever before.

It's the iterative nature of the loop that is really making things interesting. Someone might discover something new, and contributes it to a global discussion. Someone immediately takes that idea, applies new knowledge to it, and enhances it. Moments later, instant global feedback descends on the idea, morphing it, changing it, building upon it. Ideas are tossed into a global discovery engine, and are instantly transformed.

The infinite idea loop is a powerful thing – and it is changing the world faster than ever before. The pace of R&D and discovery of new ideas has forever changed within this global collaborative network. It's an eternal discussion about what comes next!

The result is that no one can hope to define the future any-more – the best you can do is simply to plug into the future that is being developed all around you and learn how to profit from it.

The globe has become one massive idea generation machine, with new ideas being created instantly, everywhere, on a continuous non-stop basis.

You do that by developing a culture that supports a highly-tuned radar that listens to the global infinite idea loop ... the future is being developed all around you, and your success comes from your ability to plug into it!

How can you do this? By participating! There are countless blogs, discussion lists, and topic websites that you can tune into. Find the ones that have the biggest impact on your career, company or industry. That's your own infinite idea loop to track.

Devote 15 –20 minutes per day to catch up on the new ideas which are emerging. As you do this, you'll find that you are constantly learning new ideas that can be applied to challenges that you face each day, and countless ways of thinking about the opportunities that are emerging in the fast paced economy.

That simple activity – learning to tune in – will provide you with insight and ideas, which are themselves the fuel for innovative thinking.

I recently spent time with a global industrial powerhouse, an organization that manufactures a tremendous amount of very sophisticated product. Over dinner, the senior VP of Marketing used a very simple example to put into perspective how the infinite idea loop was causing very high velocity change within their industry.

He described to me the scientific concept of "regenerative technology," by putting into context the idea of a "hydraulic powered bicycle."

For years, engineers in his industry, much of which involves specialized hydraulics, have undertaken research into the concept of a bicycle that can store excess energy. They've always hoped to learn how to build a bike that can capture any excess energy that is generated as it goes down a hill – using a hydraulic system as the storage mechanism. This extra energy can be accessed as the bike then goes up a hill. It's an idea that has been around for a long time, often explored, but never quite mastered.

But it could provide for breakthrough products. Such a technology, for example, could provide useful benefits for an aging population.

Here's where the power of the infinite idea loop comes in. He explained to me that years before the emergence of the Internet, the few hundred scientists worldwide who might study and think about the concept of "regenerative technology" and "hydraulic bicycles" would work in splendid isolation, with information and progress shared at a rather glacial speed. Now, however, their ideas, insights, and results from experimental projects can be shared on an almost instantaneous basis.

The result is that the science of hydraulics-based re-generative technology is now evolving faster than ever before – with the result that new product innovations with such a unique bicycle are quite likely within the next few years, if not sooner.

It's So Yesterday!

Take the "10 Things Test"

ONE OF the best ways to get a sense of the velocity that is occurring as a result of rapid scientific change and the infinite idea loop, is by taking a look at the world around you, and thinking about how it might change.

I call it the "10 Things Test."

Essentially, sit in a room, whether at work, home, in a factory, retail store or wherever you might be, and take a look around. Compile a list of ten items that you see, and then sit back and ask yourself, *"How might these things change in the next decade?"*

If you really took the time to think about the items you examine, you might be very surprised by the depth of the change that is coming. Here's what I saw with my "10 Things Test" in my home office:

- **Paint.** It turns out that "white" could be the new "green" when it comes to the world of paint. Dulux, one of the world's premiere paint manufacturers, is actively involved in learning how to use starch based plants such as potatoes and wheat to replace upwards of 25% of the petroleum based products used in a typical paint. Given the increased focus on the environment today, this could be a significant and market-leading innovation.

- **Window shades.** Think "smart-glass." Our need for window shades will soon be eclipsed by intelligent glass that will automatically adjust its opacity and transparency for various conditions. The windows will also soon be covered by a film that absorbs sunlight which will generate electrical power. Whether it's bright sunlight, a need to better manage heating and cooling costs, or to provide for greater privacy, it's likely that we'll see rapid changes with this basic component of the home and office.

- **Tissue box.** It's not the tissue itself which will have changed, but the retail technology which interacted with the box as you worked your way through the store. The box itself will have developed intelligence; it was busy updating the stores inventory system and revenue sales figures as you walked with it out the door. (You didn't have to go to a check out; they're so yesterday!)

- **Eyeglasses.** Sure, they'll still be there. But maybe they will have the ability to link directly to an implant next to the neurons in your retina, providing a direct visual link through the bifocal part of the lens for close up objects. If that's too farfetched, then a more realistic scenario would be genetic alteration of the macular tissue in your eye that would prevent any inflammatory genes from killing your vision cells – thus leading to a reduction in the leading cause of blindness in seniors – AMD (age-related macular degeneration).

- **Ceiling lights.** They'll be drawing upon the solar panels on the top of your roof and that of your neighbors. You'll have established a small community energy grid, which bypasses a need to tap into the local electrical network during the days when the sun is ready to rock and the wind is ready to roll. Solar panels are decreasing in cost at a steady pace, just as their efficiency is increasing; the same holds true for wind power. Given the likely increased volatility with traditional energy supplies, we'll see an increasing focus on alternate, micro-grid energy innovations.

The challenge in thinking about the future is that it can be difficult to comprehend the sheer velocity by which change is occurring.

- **Laptop.** What laptop? Your desk is now monitored by a 3D virtual sensor that traces the action of your fingers. You aren't really typing onto a keyboard anymore, since there isn't one. Instead, the ceiling light has directed a holographic keyboard onto your desktop; simply simulate typing anywhere with the holographic keys that you see, and your words will appear on screen.

- **Orange juice.** It will still come from Florida, but it will be packaged in such a way that the shelf life has been dramatically extended. There are huge new innovations within the world of agricultural packaging; for example, some bananas are now shipped with a special membrane that doubles the shelf-life of the product by regulating the flow of gases through the packaging.

- **Telephone.** It's likely to be "so yesterday." The next generation of kids is fully immersed in interactive tools; for them, an office with virtual 3D long distance video chat will be as normal as apple pie. Not to forget the technology behind the telephone as well; there's a good chance that you'll be sourcing your communications service from an offshore supplier, perhaps in China, Russia or South Africa. The entire industry will have defragmented and disappeared, as technological change drives many of the current business models into absolute obsolescence.

- **Eyedrops.** The trend towards hyperconnectivity will impact medical products in a big way. The packaging in which the eyedrops are purchased will "connect" to the global data grid that surrounds us, automatically pulling up a short interactive video on whatever screen that happens to be handy, with instructions on use and precautions. In effect, the role of product packaging will have been transformed from being that of a "container of product" to an intelligent tool that will help us with use of the product.

- **The view outside.** For more of us, it won't be of office towers and concrete jungles, but rather, our yards, the lake we cottage at, or the beach we play on. Ten years out, the concept of "what do you do for a living" will have changed completely to the idea of "what do you like to do?" as the itinerant career begins to dominate. (It's estimated that in just a few years, some 60% of engineering professionals will be self-employed, providing their skills on a part time basis to the global economy.) You'll be increasingly engaged in active life-design, carving out a series of activities that blend your personal interests with the need to go out and earn some funds. You'll work at a regular series of short term, highly stimulating, frequently changing project assignments. You might not have a job, but you'll certainly have some demand for your time.

Is all of this science fiction? It might seem like it, but most, if not all of the scenarios above are entirely plausible, based on science, technology and trends that exist today. A friend of mine suggested if you are having trouble taking the 10 things test, then start off with this variation: name 10 things around you that have changed in the last 10 years. Include items that didn't exist. In his case there is a laptop, a Blackberry, the iPhone, MP3 dictaphone with speech recognition, GPS unit, inflated plastic insulator packaging material, acoustic guitar with PZM mike and internal tuner, and bluetooth mouse.

The challenge in thinking about the future is that it can be difficult to comprehend the sheer velocity by which trends are occurring. That's why the "10 Things Test" can be such a valuable method of putting into perspective the velocity of change, and from that, provide a starting point to begin to crystallize some of the opportunities for innovation that surround you today.

If you examine the trends underway in my "10 Things Test," you'll find that many of them are coming about because of connectivity: eye-drop packaging, ceiling lights or a tissue box will all undergo change as they become linked into a world of massive, constant connectivity. This is a trend that has been well underway since the emergence of the Internet.

Yet this trend of extreme connectivity – or what I've been calling "hyperconnectivity" since the late 1990's – is now set to become even more pervasive, because of the emergence of the concept of "location intelligence."

Just as everyday "things" are becoming connected to the world of massive connectivity that we find ourselves in, they are also gaining the ability to advise us as to their location. This will allow for all kinds of innovative and creative new ways of thinking about "things," and how we might use them.

For example, consider what happens when industrial devices gain both connectivity, and the ability to advise us as to their location and status. During a recent session with the senior executives of a global energy company, I spent some time putting into perspective how the entire extraction and processing infrastructure of the industry will change, as every component gains intelligence and location awareness. That will lead to greater efficiency, reduced operating costs, predictive maintenance, and other potential for better operational excellence.

Location intelligence will also come to generate vast volumes of information that will be analyzed for useful patterns. Someone will figure out how to take this information and create new, billion dollar industries that involve new lines of business, challenge existing business models, or will help to take cost out of a business.

An example? After delivering a recent keynote address at a conference for a major insurance association, I discovered a fellow who calls himself a "location intelligence professional," working with the insurance industry.

He has worked to develop a career that involves determining how to link location based knowledge to existing insurance and other data, in order to revamp the insurance underwriting process. He sees a future in which up to date, location-oriented information can be used in providing for the more accurate assessment of the risk associated with underwriting a particular insurance contract.

There is an entire world of people like him who are busy marrying the concept of location-intelligence to other industries. From that, I've come to believe that we will see huge, increasing rates of change within a wide variety of industries as people let their innovative ideas take hold.

Game Time!

Can you run your business at video-game intensity?

THE THIRD pillar for the high velocity economy comes from the fact that most industries are just plain speeding up.

One way to think about this is to look at a very unique industry: the video game business. It is an industry that runs at a very rapid pace, with rapid fire product development, not to mention a certain degree of "operational intensity." Organizations that are involved in the video-game marketplace are faced with an extremely fast paced industry. They have no choice but to excel at operational excellence, given the short product life cycle times.

It's probably important to realize that the same innovation and operational intensity is coming to every industry. If this is so, then the speed at which you must innovate within your industry, as well as the intensity with which you manage your business, will have to increase.

The Chief Information Officer (CIO) of a particular video-game distributor – they operate between the video-game manufacturers and the retailers – indicates that some 45% to 60% of the total life revenue stream of a typical video game is made within the first four days of its release.

The first four days! The result is that the organization has one, simple core mission – they must ensure that new video games get to the store and get to the shelf on time.

That's why they are relentlessly, aggressively focused on operational excellence – their entire culture, information networks, management structure and organizational responsibilities are completely focused on the mission of getting the product to the market on time.

They can't afford to get things wrong; the impact would be too significant to the bottom line. And if they nail operational excellence, they nail their profits in a short period of time.

That's video-game intensity!

Could you run your business at the same high degree of intensity? At a level of operational excellence in which you are getting the right products, to the right place, at the right time, before a rapid product life cycle change renders them obsolete?

In a world of ever more rapid change, the same type of intensity is starting to come to every industry, particularly as we witness the rapid emergence of new competitors, the collapse of product life cycles, and ever increasing expectations in terms of service and support. This requires that a company think differently about its skills and capabilities; that it view its operations with a different eye; and that it develops the ability to work at an extremely fast pace.

Think about the intensity of the video game sector in the context of your industry. No doubt, accelerated innovation and rapid time to market will soon become a key trend in your industry, if it isn't already. This means that your organization will increasingly come to rely on short, sharp shocks of revenue hits, with a good chunk of total life cycle revenue happening in just a few short days.

That's why to thrive in the high-velocity economy you must focus on the concept of business intensity, and the concept of short-term, rapid operational and market excellence.

If you think about the **rapid rate of change** that surrounds **every industry** you will realize that you will soon, if not already, **be subjected to a similar fast paced marketplace.**

Remember, the key words to think about are:

- agility

- rapid time to market and

- execution.

Your market will change as it comes to be faced with the same life cycle intensity as the video game industry. To thrive in such a market, you must have agility: that is, the ability to respond to rapid shifts in the marketplace. And you've got to be able to do it extremely well – that's execution.

I spent time with a major entertainment client to help provide insight into the concept of how to be innovative within the high velocity consumer retail sector.

Certainly this particular division of the global entertainment giant is seeing their world change: new movies or TV shows now take on an instantaneity as a hit or a failure faster than ever before, which impacts any product tie-ins that this company might release to the marketplace.

The result is that the organization needs an incredible degree of agility when it comes to "time to market."

A few years ago, they had a sleeper TV movie that took off like a rocket. The company was faced with a world much different than even two years prior, as teens discussed, debated, and promoted the movie on a myriad of social networking sites. Their ability to release tie-in products to the marketplace at the speed with which the movie was "going supernova" typifies the same type of intensity as seen with the video game industry.

Take a look around any industry, and you'll find similar, customer driven intensity – which drives business intensity.

Speed Freaks

High velocity is everywhere and it drives innovation!

I T'S EASY to sit back at this point and think that your company, industry or profession won't be affected by the type of high-velocity change described so far. Video games, hit movies: sure, those industries "happen fast," but those are anomalies.

It's easy to think that way, but the reality is that high velocity is everywhere.

One of my recent innovation sessions featured a panel of three senior executives from companies that could not have been more different in terms of industry, marketplace, technology and customers: a manufacturer of board games, a heart valve company, and a sausage maker.

From the outside, it looks as though each of these organizations would be faced with different market pressures and customer demands, not to mention the rate of change imposed on them by innovation within the marketplace. We could presume that they therefore might have completely different strategies when it came to their innovation focus.

Consider, for example, the heart valve company. They are certainly impacted by the rapidity of scientific change. There are furious rates of product innovation in any aspect of the health care or life sciences industry. Their entire innovation focus has centered on how to get new products to market, which involves continuous upgrading of the knowledge of

If you look at a company, industry or career, and understand the manner by which it is being affected by velocity, you can figure out where there are opportunities for innovation.

their sales force, relentless modification of sales and product literature, and continuing education within their customer base as to the complexities of the product.

Their focus, then, is on ensuring that they are keeping up with the velocity of change occurring within their marketplace.

You might also presume that the board game maker and sausage manufacturer aren't faced with similar torrid rates of market change. And yet, they are! The board game company had recently won the rights to bring to market an edition of a popular TV game show and it had to scramble to ensure that it could pull off the feat of getting the product to market on time. One result of the velocity they were impacted by with the new game was that they were now dealing with a higher level of retailers than they had before, and hence, had to ensure they were operating at a higher level of operational excellence. This required quite a bit of innovation in terms of operating style, methodologies, information technology, as well as a complete rethinking of the information capabilities of the sales force. There was a huge amount of innovation required as they found themselves with sudden, new, unexpected circumstances.

The sausage maker is faced with their own unique sudden rates of velocity, particularly due to market success, growing revenue from $300 million to $500 million in just a few years. They found that they kept on adding staff, without really thinking through whether they were operating with enough marketplace efficiency. In addition, as they dealt with market growth, they found that they didn't have enough information to track whether the money they were spending on in-store promotions was truly effective. The growth experience meant that they weren't optimizing any of their transport planning. A quick calculation showed that many of their delivery routes were inefficient, duplicating other routes, or that trucks were sent out with less-than-full loads. The impact of the velocity they were subjected to from market growth meant that they now had some pretty serious work to do to operate at a much higher, more intelligent level.

Consider these three organizations in different circumstances and you realize there is one common issue that they have to manage: the velocity of change that is impacting them.

That is where much of their innovation emerged from. By responding to the velocity that was impacting them they were able to find some really creative solutions to challenging business problems.

It's not just business organizations that are impacted by high-velocity change. Government, not-for-profit and other organizations are also finding themselves immersed in a swirl of new issues that must be mastered.

I recently provided the opening keynote for the annual conference of a group of Chiefs of Police, speaking on the trends that would impact them in the years to come. The key part of my message? Their future success will depend, to a huge degree, on their ability to ingest new policing technologies, methodologies, staffing and skills structures.

Consider what they are up against. Police forces within the next decade will find themselves dealing with a lot of new technologies and ideas. Leading future oriented police officers throughout the world are talking of a future – within the next five to ten years – that will involve such things as unmanned aerial drone planes for surveillance and traffic monitoring; networked clothing that is linked to in-car mapping for "hot-location" information during complex police actions; and virtual reality training based on airline methodologies.

There is the issue of new methodologies and skills they will have to ensure their police staff are able to master. Increasing complexity of DNA and other forensic evidence; the increasing utilization of the Internet for crime prevention and 'crime-stopper' programs; and the challenge of using online social networking sites as evidentiary tools in complex criminal cases.

Then consider changing workforce demographics: they'll have quite a few more "retired" officers continuing to work past their official retirement date, as well as an increasingly multi-ethnic force that matches the evolving makeup of the local population.

They'll have to deal with a lot of new ideas, issues, technologies and methods. That's why Futurist magazine, in a recent column Policing the Future: Law Enforcement's New Challenges, noted that "…better-educated police officers with improved people skills and a stronger grasp on emerging technologies will be crucial to successful policing in the future …"

The key point: velocity is everywhere, and everyone must deal with it.

Hero to Zero

Is your brand from the olden days?

I N THE high velocity economy, here's another important issue that you need to think about: a brand today can go from hero to zero in a matter of months. There's opportunity that comes from innovation in going after those brands which are aging at a hurried rate; there's threat if you don't provide leadership to ensure you avoid such a fate.

Consider, for example, the fate that has befallen Sony in the last few years. The organization once had a really cool brand name. The Sony Walkman had deep brand value, yet Sony seemed to lose its innovative spirit. It ended up destroying a good chunk of the brand value behind the Sony name. If you think of the Sony name now, you might be thinking of an organization that is slow, behind the times, ponderous. The saddest thing is that Sony has messed up in so many ways, that some customers now look at them as if they have an "L" on their forehead.

The rate at which the Sony brand lost its value is nothing short of stunning but was due to a series of well known mistakes, some of which include:

- they failed to keep up with the rapid growth and demand for flat panel TV's and other hot new technologies: they failed with market agility;

- they decided that going to war with their customers in order to prevent music piracy (by slipping destructive software onto their music CD's) was more important than developing great technology that caught the next wave of consumer electronics, particularly with MP3 players;

- they dropped the ball on the necessity for continuous operational excellence, as evidenced by a disastrous recall of laptop batteries.

Bottom line? Sony lost its shine. Kids think iPods are cool, and don't really know what a Walkman is. That's an example of where a brand simply became old, perhaps before its time.

How can you assess if your own brand is falling out of date and if your brand looks tired? There are several things to watch for:

- **You are out of tune with your customers.** Case in point – many companies in the automobile industry missed out on the entertainment revolution in the passenger compartment, because they weren't watching what their customers were doing. They were busy releasing automobiles that were some five years behind the living rooms of their customers.

- **Customers see a lack of innovation.** Consumers today are immersed in a global cloud of new ideas. They're witnessing constant, relentless, awe-inspiring forms of innovation all around them, as they deal with a flood of new consumer technologies, packaging based product innovation, and ongoing advancements in retail environments, both offline and online. They've come to expect that the brands they deal with are at the leading edge in design, functionality, message and purpose.

- **Lousy, ineffective customer service.** Guess what – when it comes to interaction with your customers, they measure you up against the world's best. If you don't add up, you are doing some significant damage to your brand equity. Customer support is no longer good enough – customer support that beats the competition is the bare minimum.

Consumers today are immersed in a global cloud of new ideas. They're witnessing constant, relentless, awe-inspiring forms of innovation all around them.

- **You don't know that your customers know more about your brand than you do.** Your customers today are immersed in the infinite idea loop. They are busy sharing ideas on what's really cool, and they are even busier taking apart the folly of those who have been left behind. In doing so, they are rapidly reinventing products, services, brands and image. If you aren't listening, you are guaranteeing that you'll fall behind.

- **A lack of purpose or urgency.** Many organizations still lack the critical insight that they should have in order to provide for market agility. They don't have instant feedback mechanisms which tell them of rapid developments in specific markets. They don't know how to regroup quickly "when bad things happen." They still operate blind, as their sales force goes into a customer meeting, oblivious to what that customer has been thinking about them. They approach every day as if it were the same as yesterday; meanwhile, their market and their customers have run away from them!

- **A lack of market and competitive intelligence.** It's the information-age, get it? There's no shortage of information to be had. Yet there are companies who seem shocked when a competitor drops a 'bombshell' announcement – only to realize that they were the only one who thought it was a bombshell. Everybody else knew what the competition was up to, because in this new hyper-connected world, everyone knows what everyone else is doing!

Any one of these may seem like a small mistake, however, one mistake can be instantly compounded; a small PR mistake, a lousy executive decision, or poor execution, can lead to instant, global brand devaluation.

Avoid making a regular series of fumbling missteps and recognize that brand longevity is a critical issue. Consider these steps:

- Ensure your sales, marketing, development and customer support teams are relentlessly focused on the currency of the brand.

- Make sure that continuous brand innovation is part of your corporate mantra.

- When confronted with the new and challenging customer, learn from them rather than running away.

- Be incessantly focused on the need to constantly evolve and update your product and service at the rate the market demands.

- Make sure that you track the new emerging technologies, scientific advances, consumer attitudes and all the other factors that could influence your brand development in the years to come, so that you don't get 'surprised'.

- Learn to think five to six product lifecycles in advance and plan to do them all within six months.

- Make forward oriented intelligence a critical aspect of what you do.

Part of the challenge today for any existing brand is that anything you do – or don't do – can immediately be examined within the global microscope that is the Internet.

Consider what happened with WalMart when it established WalMart HUB, an online site that was supposed to draw young kids who might otherwise be spending time on the popular social network site, MySpace. Within a matter of days, the site was the talk of the online chatter circuit, being ridiculed by the very audience it was trying to attract. Not only that, but the advertising media had taken the site apart, examining the missteps, and ridiculing WalMart for looking so dreadfully uncool. The site referred to members as "Hubsters" and had teen comments such as "Fashion is like my #1 thing!"

What went wrong? Certainly WalMart was being innovative, it was trying to utilize new emerging technologies to appeal to the rapidly changing activities of its customer base. But it didn't quite understand that everything it did was completely opposite to what its customer base was actually doing online on MySpace.

As the Guardian (UK) newspaper noted, WalMart had it all wrong with the overall concept of the site: "Any teenagers wishing to sign up as "hubsters" need their parents' consent and entrants face the challenge of looking cool in Wal-Mart apparel: videos and web pages are banned from carrying trademarks, trade names, logos or copyrighted music. However, clothing bearing the Wal-Mart stamp is allowed." WalMart took the site down within a matter of days.

Lesson learned? Innovation is critical: yet, you must always approach creativity and innovation with an understanding that your every move will be analyzed, and instantly subjected to a global up or down vote.

Ready, Set, Done. • How to Innovate When Faster is the New Fast

Time Travel

Can you deal with velocity?

I T'S QUITE easy to lose sight of the massive rates of change occurring within every industry, profession, business model and product / service model. Yet it's by riding the crest of change that many opportunities for innovation can be found.

So how can you ensure you keep up with high-velocity change? Take the time to assess the rate of change that is impacting you, your company, industry or profession. But first, get in the right frame of mind.

Here's a neat trick – go dig out your old marketing textbook, and look at the chapter on product lifecycles. I dug mine out, and it showed the product lifecycle as being something that lasts for 15 to 20 years!

Sit back and think about just how ridiculously old-fashioned such a concept is. Given the rapid product development and faster obsolescence, in many industries there are no product lifecycles today. Given that, there is just "agility in time to market." Success is not defined by how long a product will last in the marketplace, but by how quickly you can get a new product out there before it's out of date.

Do you have that type of agility? Maybe not: that's why it is important to link the concept of "agility" to "velocity." When thinking about innovation and the future, it's your ability to

Success is not defined by how long a product will last in the marketplace, **but by how** quickly you can get a new product out there before it's out of date.

stay ahead of the innovation curve when it comes to what you do, and how you do it.

Ask yourself these questions to determine if you are truly prepared for velocity:

- Do you know the rate of change – the velocity of change – that impacts you, your industry, your products?

- Are you meeting the requirement for operational excellence that your customers, suppliers, business partners and everyone else expects of you, in a world in which such expectations are constantly changing?

- Are you properly positioned for velocity, in terms of your ability to do things at the pace required?

- Does your culture support high-velocity change, or are you almost keeling over from organizational sclerosis?

- Do you have the feedback and innovation mechanisms in place to deal with high-velocity change, and can you learn from them?

- Are you planning at the leading edge, or are you still reviewing what you were planning last week?

- Are you evolving markets/products at the pace required, or are your customers marching on because you are stuck in a slow time-to-market rut?

- Are you at the curve of expectations of customers needs – do they think you've got the right stuff when it comes to velocity?

- Are you anticipating customer solutions before they know they need them?

These are all questions that can help you assess the velocity that you are being subjected to, and can assist you in identifying some of the areas where you might need to light an innovative spark.

Here's an important thing to realize about velocity: change that can impact your industry will often come from outside your industry, and if you don't have a great 'trends radar,' you'll miss the velocity of change.

Consider the auto industry. For years, they didn't tune into the fact that consumers were increasingly migrating to MP3 players, iPods and other forms of music entertainment accessories. The result was, as one joker put it, "my car is some ten years behind my living room."

Only belatedly did they realize what was going on, at which point they began to add new capabilities to the automotive entertainment system, such as iPod docks.

Sadly, this lack of insight into consumer-driven velocity seems to be a hallmark of Detroit auto companies. Witness how they often miss the signs of a customer switch from big-engine vehicles, to small, fuel efficient hybrids.

Critical to understanding your future is the velocity of change that your customers subject you to. In many industries, customers are defining the rate of velocity, and it's by watching them that you can best understand how you will be impacted.

Ready, Set, Done. • How to Innovate When Faster is the New Fast

Agility

Agility

Why solving the skills challenge will be critical to your innovation capabilities

HIGH VELOCITY change demands innovation – that much seems clear. So shouldn't you just be able to set out, and figure out how to do things differently in order to keep up with our fast paced world?

If only it were so easy. The reality is that your ability to innovate will be greatly influenced by how well you do at marshalling, attracting, and educating the skills that you will deploy to meet the challenges and opportunities of our fast paced world.

Here's why:

- Success in the future will come not only from your ability to generate innovative and creative solutions to problems, but also from your ability to deploy the right resources at the right time to deal with those challenges. In "**Talent Agents,**" we describe the concept of skills agility, and put into perspective why it is going to be a critical success factor in the future.

- Given the importance of skills agility, you also need to understand that it is going to become increasingly difficult to achieve, an issue that we outline in "**Speedy Skills**." That's because the infinite idea loop is leading not only to faster innovation, but very rapid growth in global knowledge, in almost every single industry, profession and field of human endeavour.

- That's leading to a new reality – your ability to attract, retain, engage and utilize specialized skills will be a critical success factor. That's why you need to understand the "10 unique characteristics of 21st century skills" that are outlined in "**Skillful!**"

- To put further clarity on the uniqueness of skills, a key issue is that skills are becoming extremely fragmented and specialized, to such a degree that entire new professions are emerging all around us. In "**Innovation Fertilizer,**" we outline a perfect example of a unique new career – manure managers.

- Put together the unique trends occurring with knowledge and skills, and a new reality emerges. Your ability to innovate will come to depend on your ability to generate new knowledge at the pace demanded in the high velocity world. In "**Fast Facts**," we provide details on the concept of "just-in-time knowledge," a critical capability that will be crucial to your innovation efforts.

Talent Agents

The importance of skills agility

O NE OF the defining success factors in the high velocity economy involves your ability to tackle complex problems with the right skills base.

In an era of rapid innovation, you are faced with a future that requires a new form of human capital: the ability to deploy the right skills at the right time for the right purpose, regardless of where the skill might be required or where the skill is sourced.

Clearly, the ability to access skills will be a critical battlefront in the future. We're witnessing the rapid obsolescence of skills as a result of the rapid emergence of new knowledge; we are seeing constant career upheaval and in some cases, the disappearance of careers; we're already seeing a shortage of critical skills throughout almost every industry sector, particularly with skills that involve science based knowledge.

The impact of these trends is compounded by significant demographic changes and workplace attitudes that will make it more difficult to find the right skills that are needed for an increasingly complex world.

Given the fragility of the existing skills that you already have, getting access to the right skills in the future might become an even more significant challenge! That's why focusing quite a bit of your innovation energies on the concept of "skills agility" will be critical. Already, there are signs that some of the world's market leaders are focused on the concept of "human capital agility" – or what we might call "just in time skills deployment."

Whether it is dealing with the rapid emergence of new client expectations, ever more challenging organizational complexities, more complex product development, or sudden and dramatic new forms of competition, they are orienting themselves toward the ability to quickly harness together a new team of talented people. That capability is becoming a key critical success factor for the future.

This is an area that is ripe for innovative thinking. From an innovation perspective, you've got to constantly assess whether you've got the depth and scope of skills that you might require as the world goes high velocity. This involves constant, probing questions that continually assess the organization and its skills. Questions such as:

- do we have the people we need in the right places/positions to fulfill the mission?

- do we have the right people with the right skills available at the right time?

- if we are suddenly faced with rapid market or industry change, do we know how to access specialized skills and talent we might need?

- what barriers might exist in our ability to rapidly acquire and deploy skills?

A big part of the innovation opportunity comes with the global connectivity that has emerged over the last few decades. There is plenty of opportunity for an organization to do what couldn't have been done even five years ago when it comes to skills accessibility. Today's workforce can be accessed and harnessed in ways that were not previously possible; unique and specialized skills can be sourced from anywhere in the world.

From an innovation perspective, you've got to constantly assess whether you've got the depth and scope of skills that you might require as the world goes high velocity.

What is the ultimate goal of skills agility? It's keeping up with the concept of "high-velocity" to help you to:

- provide for business flexibility in a time of rapid change;

- avoid higher recruitment costs through imaginative struc-ture / relationships;

- shift the balance of the current supply vs. demand skills imbalance by changing the terms of the current skills imbalance;

- access an increasingly global talent base.

When you witness an organization that has real skills agility, it can be an awesome thing to study and understand.

A recent client of mine was the Association of Organ Procurement Organizations, representing the teams of health care professionals and volunteers who are involved in helping families understand the benefits of organ donation, as well as the medical and surgical teams who are involved in the transplant process.

Sadly, they often have to kick into high gear during extremely tragic circumstances involving the sudden death an individual, most often involving serious head injuries, which result in the potential availability of organs for donations.

But talk about agility! These teams quickly self-assemble and self-organize based on a long track record of providing instant response to what unfolds as an extremely complex process. A part of the team discusses with the family the benefits, process and need for organ donation. At the same time, when it appears a donor might be likely, an extremely sophisticated, finely tuned process kicks into gear, in which a transplant team is notified. A logistics network procures emergency air transport, when needed, and other forms of transport. Multiple different hospitals are immediately brought into the process, to ensure that surgical rooms are set up and ready to go.

All of this unfolds within a matter of hours; timing is absolutely critical.

The fascinating thing about this group of people, who come from all across the country, is that they consist of a wide variety of skills backgrounds: highly trained heart, lung and other surgeons; nurses or other medical professionals who have developed unique counselling capabilities; volunteers who provide their time to both the counselling process as well as management of the logistics of the entire process.

If you think about the complexity of what these teams manage to accomplish each and every day, with very little advance planning, then you've established some great insight into the concept of skills agility.

Speedy Skills

The impact of information and knowledge on careers

I F SKILLS agility is important to your ability to innovate in the high velocity economy, then you had better also understand that achieving such agility is going to become increasingly difficult.

That's because another key impact of the infinite idea loop is that the volume of information, and resultant knowledge, is ballooning at an ever increasing pace. As this happens, the complexity and availability of specific skills will become more difficult to access.

According to a study out of the University of California (Berkeley), we now produce as much new information every six months as was produced in the first 300,000 years of human existence.

One result of this trend is that every profession, job and career is evolving and changing at a rapid pace. Years ago, Futurist magazine observed that

> "The pace of technical change is so fast now that we must be prepared for a man to change not only his job, but his entire skills, three or four times in a lifetime."

Consider the impact another way. Australia's national government did a study on future career trends. In the national government report the Innovation Council Chairman indicated that 65% of the children who are in pre-school today will work in jobs and careers that have yet to be defined.

What is the impact of such trends on careers? There are several:

- **It speeds up the volume of knowledge that people need to know to do their jobs.** All of the knowledge we have today will be but 1% of what we will have in 2050. It is estimated that medical knowledge is doubling every eight years, and that half of what students learn in their freshman year about science and technology is obsolete or revised by their senior year. Consider this – one professional estimate suggests that the 1/2 life of an engineer's knowledge is about 5 years. Clearly, people are now expected to master more knowledge than ever before, at a far more rapid pace.

- **It speeds up the pace of research and development, which leads to an ever-increasing rate of change with the products and services that individuals are responsible for within their career.** We now have, with the global connectivity that has emerged, a global petri-dish for knowledge exchange. That means that the pace of technological change now accelerates with each new discovery, leading to an ever-increasing pace of new development. The result? Constant innovation is critical in every industry and profession in order to stay in the game.

- **It helps to foster greater competition in every marketplace.** A faster rate of R&D, which results from increased global information sharing and better corporate intelligence, means that as products come to the market quicker, the very nature of competition in many an industry is heightened. This helps lead to the greater degree of structural, ongoing corporate change that we see around us today. As competition increases, organizations are forced to undertake bold new actions to retain market share. They must constantly attack their cost structure, seeking to squeeze out new efficiencies and cost savings. They relentlessly examine and pursue new external

While some individuals find that their chosen **profession is undergoing a constant evolution** into multiple specialties, others are finding that **they must master multiple different specialties.**

partnerships with the aim of outsourcing as many ongoing functions and responsibilities as they can.

- **It changes the very nature of the professional or degree skills that many people possess, and leads to new day-to-day responsibilities.** As new products and services come to market, the very nature of the day-to-day activities of everyone undergoes rapid change. The most fascinating thing is that it makes it impossible for anyone to cope with everything in a particular profession or skills area. The result is an increased degree of specialization within what were once broad areas of knowledge. For example, while in the past we had "human resource professionals" who could handle a wide variety of issues, the field of employee and career management has become far too complex for any one individual to manage on their own. The result is that today we have career counseling specialists, as well as people who focus on training and development, others who are specialists in corporate use of instructional technologies, and others who spend their time on issues of labor or industrial relations. Not to mention pension and group benefit managers, as well as career planning experts. Every profession, every career, and every job is being sliced and diced into multiple sub-categories.

- **It leads to an increased need for "multi-skilling."** At the same time that we are witnessing ever increasing career specialization, we are also finding a need within many professions for people to take on multiple different roles. While some individuals find that their chosen profession is undergoing a constant evolution into multiple specialties, others are finding that they must master multiple different specialties. They must be able to take on a wider variety of roles and responsibilities, and are expected to pick up an ever-increasing number of new skills.

- **A demand for "just in time skills."** One of the major impacts of our world today is that the rate of change has become nothing less than dramatic. People are now expected to be able to master new topics and issues at the drop of a hat. Nowhere was this more evident than with the emergence of SARS in 2003. The health care professionals discovered that they needed to quickly learn about a wide variety of issues. Many medical associations found themselves scrambling to put together educational conferences that focused on the many different aspects of SARS, ranging from medical education, to crisis management to issues of public safety and education. We have never seen a medical world evolve so rapidly. We went from having literally no information about this new disease, to a situation in which it is estimated that the collected global information available on SARS alone exceeds the extent of global medical knowledge that existed in 1965. We can expect this type of knowledge demand to become more frequent in the future. A world of rapid knowledge advance implies that a greater number of people will be expected to learn instantly about new topics, issues, products and strategies.

A world of constant new knowledge growth implies a world of constant change – and constant change demands regular knowledge transfer, upgrading and development. That's why one of the best phrases to focus upon is this one: "learning is what most adults will do for a living in the 21st century."

Companies in the computer and consumer technology industries are at the vanguard of the knowledge challenges that organizations must deal with in the future.

One recent client, based in Silicon Valley, indicated that because of rapid innovation, driven of course by rapid science, they are now witnessing some products with lifecycles as short as three months.

They came to realize that there was a key opportunity for innovation, by putting in place a sophisticated, technology based "knowledge replenishment system."

Designed for their sales, customer support, marketing, and most other staff, it was put together to ensure that everyone is kept up to date, not only with the features and capabilities of the current generation of product, but that they are also in tune with what comes next, and what comes after that!

In this way, they could be certain that they could focus on time-to-market, be in tune with the rapid turnover of their product base, and be fully aware of how to sell, market, support, and manage a continuous flood of new products.

Skillful!

10 unique characteristics of 21st century skills

GIVEN THE importance of skills agility, will it be an easy problem to solve? Maybe not. What is certain is that you'll need plenty of innovative thinking to deal with it.

While there has been a lot of talk about the skills crisis lately, most of it is focused on the wrong thing. People seem most worried by the fact that a lot of baby boomers are set to retire, and are taking their skills out of the economy. That certainly is an issue, but that's not the big issue.

The big issue is if you don't understand the uniqueness of 21st century skills you won't be able to access any.

There are several major workforce trends that will impact you in the future:

- every organization is faced with an increasingly complex, restless, age-diverse, disloyal, and highly specialized workforce – a workforce that will have the longest life-span ever, from hyperactive 15 year olds to wizened, not-ready-to-quit 80 year olds;

- with the coming "end of retirement," most companies will come to realize they'll need a lot of telephones with big buttons for the 70+ folks who are still a part of their work-force – and a lot of innovative workplace practices as well;

- the arrival of "Gen-Connect" (the kids who have been wired with a mouse since birth) will lead to the question of whether "good luck" will be the only possible response to the issue of managing this generation;

- this workplace weirdness will only be compounded by the ongoing rapid evolution of knowledge and skills, such that most organizations will find it impossible to find the highly specialized skills needed in the economy of the future.

The "War for Talent" will be the new competitive battleground. Organizations that can attract, engage, retain and amuse an increasingly complex workforce will be the ones who find success in the rapidly evolving global economy.

In an era such as this, firms are faced with a future that requires a new form of human capital agility: the ability to deploy the right skills at the right time for the right purpose, regardless of where the skill might be required, or where the skill is sourced. At the same time, organizations are faced with an increasingly global talent base, a reality that demands new forms of collaboration, insightful project management, and deep insight into the effective utilization of those skills. The way to the future is clear: it's no longer about managing time: it's about successful skills deployment.

If an organization is to survive the high-velocity economy, it needs to be doing a lot of innovation with how it deals with skills, and it needs to understand the key attributes of skills in the 21st century:

- **Skills are more specialized.** Rapid knowledge growth means that it is increasingly difficult for people to keep on top of what they need to know. That means people need to specialize; knowledge niches are the reality for most professions and careers. As they specialize, simple supply/demand reduces skills availability, leading to skills inflation. It's going to cost more to get the right specialized skills – that's a big problem.

- **Skills are disloyal.** A recent survey out of Belfast indicated that 36% of people, on their very first day on a new job, were already thinking about looking for another job! That's probably not unique to the Irish; and if so, it confirms that a massive philosophical shift towards a "job" and "career" is underway. The death of corporate loyalty means an increasing difficulty to get the right skills.

- **Skills are degradable.** The half life of knowledge is decreasing at a furious rate. Most organizations are discovering that the skills they do have are becoming increasingly useless as knowledge obsolescence takes hold. Skills are ready to walk out the door as soon as they arrive – and if they hang around, their value decreases rather quickly!

- **Skills are renewable.** Fortunately, out-of-date skills can be given new life. If people and companies can develop the ability to generate just-in-time-knowledge they'll learn how to adapt and evolve.

- **Skills can be complacent.** The challenge is that a lot of organizations don't really worry about the points above. Some professions, and many staff in organizations, simply don't think about the reality of skills extinction as a real trend. They have no desire to upgrade, enhance, or change their capabilities. The lack of urgency leads to a sclerosis that impacts the overall ability of the organization to change, innovate and create.

- **Skills are disposable.** The unique thing about skills today is that companies clearly don't need staff anymore – they simply need the right skills at the right time for the right purpose. After that need has gone, they will need different skills for a different purpose. In the high-velocity economy, the idea of a permanent skills base is a quaint concept from the 20th century.

- **Skills are increasingly portable.** That's the good thing we've learned with globalization: with the depth of the

The "War for Talent" will be the new competitive battleground.

emerging skills crisis, it doesn't really matter anymore where the skills are – as long as you can get them, that's all that counts!

- **Skills can be transferable.** The boomer retirement issue is real. Smart organizations are spending big money to ensure that important knowledge is captured, retained and archived.

- **Skills should be experiential.** The concept of skills or experiential capital is one of the most important assets that a company requires. It's the combined knowledge that has been learned through innovation, risk, failure and success. Boost that skills capability and you've done something that flows onto the bottom line.

- **Skills are generational.** We're going to have a lot of active 80 year olds in the economy as the end of the concept of retirement draws near, at the same time that companies seek skills from bright, knowledge aggressive 15 year olds. We are going to have the longest life-span economy that has ever existed. If we prepare for that culturally and organizationally, we've got a good strong plan for dealing with the skills challenges of the future.

In the high velocity economy, talent, not money, will be the new corporate battlefront. Managing and obtaining skills is going to require a lot of innovation and creativity in terms of solutions.

My perspective on the future of the skills issue was instantly determined when I read a New York Times editorial in October 1987, "Tomorrow's Company Won't Have Walls." I had already been immersed in the nascent world of online communication for some five years, and had come to believe that the ability to network with knowledge would come to have a big impact on the future. Yet, I still wasn't quite clear in my mind as to what it all meant.

The editorial put the future into perspective in concise, crystal clear terms, noting that we would see the emergence of a "network organization" that would draw together necessary skills from throughout the world.

Such an organization would scale up and scale down as necessary, taking on new projects, and defining its very success through its ability to access skills: "… competitive advantage will rest increasingly in the way each network organization gathers and assesses information, makes its decisions and then carries out those decisions."

The article also commented on the temporary nature of the future skills relationship, noting that "people from the network and from outside the company will join the group at the hub for periods of time and then leave it."

I never looked at an organization in the same away again. Having said that, I still believe that few organizations have mastered the concepts found in the editorial, and that there remains a huge opportunity for skills innovation.

Innovation Fertilizer

Find signs of
the future in manure!

ARE 21ST century skills really going to become all that complicated? Are you really going to find yourself in a situation in which your ability to innovate at the pace which is required will be challenged by an inability to attract necessary skills?

Yes! And you can understand this in a greater degree of depth if you examine where we are headed in the future by taking a look at what is going on with manure.

Manure?

That's right – manure.

The potential of what can be done with manure today has expanded significantly through the last several years, to such a degree that the typical farmer can no longer be expected to know everything there is to know about manure. There is an entire industry of folks who possess the specialized knowledge, innovative mindset and deep insight into the world of manure and how it can be managed, something that is of critical importance to the farming industry.

Why has manure become complex? Let's look at one of the biggest manure management problems that farmers have – the "pit crust." As the name suggests, this is the top layer of the manure in the pit, and it gets rather hard and crusty. All farms have a pit – something has to be done with the waste generated by the animals – the thicker the crust, the greater the number of flies and rodents, not to mention enhanced smell problems.

Enter the world of bio-genetics. Scientists at a bio-genetics company have learned, through detailed research, that most of the pit crust on the manure comes from the outer shell of the corn that is fed to the animals. Their solution was to develop a specialized bio-enzyme that would break down the corn shell during the digestion process, with the result that the pit crust could become dramatically thinner. The result has been fewer rodents and flies, less potential for disease, and a big, positive environmental impact.

Bio-science is affecting agriculture in a huge way, as are many other aspects of science. The result is that what farmers need to know to better manage some of the unique problems on their farm is increasing exponentially. The pit crust problem and potential solution is but one small area of science that is part of a larger picture. Given that there is so much new stuff going on, the typical farmer might not become aware of this type of product.

That's where the 'Manure Manager' comes in. The Manure Manager is a partner to the farmer, providing them the very specialized, niche oriented knowledge that they have developed. They are individuals who possess the specialized skills of knowing "what's out there," and "what can be done with it." They are partners to the process, helping the farmers cope with the rapid rate of change that is swirling around them.

These manure experts are also figuring out how to provide farmers with significant revenue enhancements, through the more intelligent application of manure on the fields. Manure is often used as a nutrient on the farm, and farmers deal with very complex methods of determining how much nutrient is needed for each particular crop, on a certain type of soil, in a certain location.

That's where we are seeing some of the most far-reaching aspects of manure management. In one area in the Midwest, a group of manure management experts have been working with local farmers to undertake very detailed soil and yield analysis, to determine the best application rates for future plantings. The returns have been significant – one family farm saw a $19 per acre increase in revenue yield through such efforts. That might seem like a small number until you multiply it by 2,000 acres, for a net result of $38,000, a big revenue improvement for a family farm operation.

This is part of a larger trend called "precision farming," which involves a very sophisticated marriage of leading edge technologies with the planting process. In precision farming, a farm tractor uses GPS receivers to link to a satellite in order to determine its exact position. It then uses sophisticated yield analysis techniques, often with the involvement of the manure management consultant, to determine exactly what needs to go in the ground in terms of seed and fertilizer, in order to produce a better crop at that exact spot.

Then there are farmers who are turning manure into money! In Vermont, there's a program underway in which they are innovating with a variety of methods to turn manure into electricity. The incentive? The farmers are paid four cents a kilowatt hour above the market rate, as part of a state drive to find alternative energy sources. They are doing so with the help of engineering and electrical specialists – and local manure managers who understand everything they need to know to deal with this unique fuel-source substance and maximize its output.

What do these "manure managers" really do? They excel in focusing on a very specific niche. They understand the rapid evolution of knowledge, methodology and innovation occurring in that field, and make their expertise available to a huge industry.

There is so much new knowledge emerging around us, every profession and career is fragmenting into hundreds and thousands of sub-specialties. No one individual can be expected to master everything anymore; instead, they'll be responsible for some type of core responsibility (as is a farmer) and will

No one individual can be expected to master everything anymore; instead, they'll be responsible for some type of core responsibility and will spend their time working with the required subject experts.

spend their time working with the required subject experts (such as manure managers, among many other specialists).

We are becoming a world subdivided into the generalists, and the specialists. Some very innovative thinking can be found as to the nature of partnerships, the rapid evolution of science, and what organizations need to do to partner up with very specialized skill sets that are emerging.

The ability to lock up access to critical, specialized skills will likely become a key method of achieving a competitive advantage in the future.

I was invited to speak at a financial industry conference held on the island nation of Grand Cayman. One of the issues that I focused on was how quickly financial skills are becoming specialized. I raised the question as to whether the nation that could "lock up access" to these skills might find the key to opportunity for the future.

Certainly the high-velocity change is there. New business models in the financial services sector are emerging faster than ever before. There is an increasing volatility with global money, such as that seen with the global rush to private equity. As the financial services world becomes faster and more complex, the skill set of those involved in high end banking services is becoming ever more precious and scarce.

Already, the signs are there, in that offshoring within the industry is now going strategic: it's not just about saving money any more. Noted a recent article in Asian Banker: "… outsourcing will become less about cost containment and more about accessing the best skills and expertise…"

In other words, one of the key ways to establish a competitive advantage in the global financial industry will come from getting access to extremely specialized skills: if you can get the right skills at the right time for the right purpose in a financial marketplace, you might survive the challenges of the future, or you might find the key to competitive advantage.

That's where innovation comes in: the trick to thriving is riding these trends to success, by being open to change, by innovating with skills access: thus recognizing skills accessibility capital as one of the key economic success factors of the 21st century.

Fast Facts

Just in time knowledge

I F SKILLS agility is critical, and yet skills are evolving at such a furious pace that a shortage looms, how can you ensure that you can keep your innovation capabilities going?

By focusing on the concept of "just-in-time knowledge" you ensure that the people and partners involved in your organization, can pick up the right knowledge, at the right time, for the right purpose.

Whether we are dealing with medical, scientific, financial, business, mechanical or engineering issues, one thing is clear: the knowledge that you need to know to do your job today is becoming infinitely more complex every minute, with a constant, relentless flood of that which is new. In such an environment:

- the ability to obtain rapid, instant knowledge is becoming an urgent necessity in almost every field of endeavor;

- the ability to quickly digest, understand and assess new knowledge is an increasingly important skill – one that not a lot of individuals have mastered;

- the ability to reformulate our thinking, assumptions and capabilities to respond to the constant change being thrust upon us is of increasing importance.

That's where the concept of "just-in-time knowledge" comes in, as it best describes the nexus of these realities. From an innovation perspective, there is plenty of opportunity for meeting the demands of our fast-paced world through just-in-time knowledge.

Just-in-time knowledge isn't the same as "continual learning." That particular phrase became popular when it became evident that all of us would have to constantly refresh our knowledge base in order to keep up with change. We've done a good job at mastering that challenge, and many people have made continual learning a key part of their career path.

It's time to go to the next step, and assess how you can develop the capability to generate "just-in-time knowledge," a type of knowledge that is one step beyond continual learning. Indeed, it's a form of continuous learning that is instant, fast, and urgent. Think about situations where a need for just-in-time knowledge is evident:

- Some estimates suggest that medical knowledge is now doubling every eight years. Rapid advances in new methodologies, technologies, treatments and methods of care evolve at a furious pace. In such a world, medical professionals can't be expected to know everything there is to know within their particular field of endeavor. The new reality going forward for doctors, nurses and any other professional is that these professionals are increasingly forced to go out and obtain new knowledge, just at the time that they need it.

- Sales based organizations are quickly discovering that furious rates of hyper-innovation in their marketplace require a sales force that is extremely adaptable, agile, flexible – and quick to understand the potential of new markets. If a product has a life of about six months in the marketplace, an organization can't afford to waste any time in preparing to assault the market. The result is that there is an ever increasing need for sales based organizations to gain deep, rapid insight into the sales potential of

Just-in-time knowledge: a form of continuous learning that is instant, fast and urgent. The right knowledge at the right time for the right purpose for the right strategy.

a new product line, while discarding the knowledge and understanding they have of the old product line.

- Mechanical engineers continue to see rapid developments in manufacturing methodologies, as well as a need to quickly master the art of offshore manufacturing. Rapid technological change means that every engineer involved in process automation must have the ability to quickly gain insight and intelligence into leading edge issues associated with plant design, construction, automation, assembly, robotics, and all kinds of other complex topics.

In all of these cases, the individual must be able to quickly obtain the right knowledge at the right time for the right purpose for the right strategy, all revolving around the fact that the knowledge around them is instant, fast, and transitory.

In order to keep up with the fast change that results from fast knowledge evolution, organizations and the people within them must focus on developing the capabilities to grab that knowledge instantly and at the right time. That involves a huge opportunity for innovative thinking – all kinds of new methods exist for internal and external collaboration, information sharing, knowledge capture and distribution, and countless other methods of knowledge enhancement.

A great example of an industry that has to deal with just-in-time knowledge involves the global credit card industry, which is being impacted by the rapid emergence of new technologies. I was immersed in this industry when invited to speak to one of the companies that manufactures the highly specialized credit card embossing technology.

What's the industry faced with? Consider, for example, the concept of "contactless payment technologies." We are witnessing the rapid emergence of a wide variety of payment methodologies which embed the credit card or other financial information in various devices, such as cellphones and key fobs.

An industry newsletter, Card Tech, commented that things were changing so fast that "our forecast, built a few months back, underestimates the market, there's been so much takeup. It's been explosive…"

Financial institutions, as well as the companies that manufacturer the infrastructure of our credit card world (including credit card embossers, gas pump manufacturers, debit/credit card machines), are all having to deal with the fact that there are many new ideas, and hence, many new and different technologies that provide for contactless payment.

This means that their management, sales teams, R&D groups, and just about everyone else has to be in a state of constant readiness for rapid change, doing so through constant, relentless learning about the new ideas emerging around them.

In other words, learning about the constant change occurring within their industry, just-in-time.

Innovation

Innovation

It needn't be difficult to do!

HIGH VELOCITY demands that we do things differently in order to keep up with an extremely fast paced world. That's what innovation is all about. But where do you start? That's what we'll cover in this section.

- Many people seem to remain mystified by the concept of innovation. Years of media coverage have led to a belief that it's all about bringing new products to market, and that much of it is related to the design that goes with new products. Yet that isn't necessarily so. In "**Simply Uncomplicated**", we outline the three essential elements to innovation: running the business better, growing the business, and transforming the business.

- Organizations that get beyond the mystery of innovation learn something new: they champion different ideas, welcome the sharing of ideas, and are inspired by an open culture. Those are just a few of the elements of "**Innovation Oxygen**" that you can breathe to get underway with your own innovation efforts.

- But innovation doesn't happen magically, you can't just will it to occur. You need an organizational culture and attitude where many people move beyond day to day management, into a role where they are all helping to lead the organization into the future. That's why "**Navigating Innovation**" is critical – to build an innovative organization, you need an organization wide team

Innovative ideas come from observing the trends that will impact you, and thinking about what you should be doing to respond to or stay ahead of those trends.

that is focused on leading the organization into the future.

- There are certain goals that innovative organizations pursue: cost excellence, a growth orientation, and time to market. There's much more. In "**Step Forward**", we examine the nine key attributes that are shared by the innovative organizations that we've studied.

- How can you put some clarity on your innovation goals? Undertake a "**Pinpoint Analysis.**" It's a way of thinking about the current state of your world, markets, business and structure, to identify all the areas for potential improvement.

- Certainly undertaking a critical self-examination can help you identify opportunities for innovation, but don't stop there. In every single industry, there are always all kinds of other offbeat sources of ideas. In "**Rebels Yell!**" we take a look at some of the places you should be looking for other inspirational insight.

- Have you ever thought of looking to truckers for inspirational insight? You won't find many of them out on the speakers circuit sharing their unique innovative methods, but they are innovation leaders. In "**Keep on truckin'**", we look at the secrets to their innovation success.

- Innovative ideas come from observing the trends that will impact you, and thinking about what you should be doing to respond to or stay ahead of those trends. Quite often, the first step is not done well: most people think only about short term trends, and not the trends of real substance. That's why "**The Word Is: Transformation"** will help you think about the real, significant, transformative trends that will shape our world in the years to come.

- You'd also do well to spend some time to find some "**Innovation Heroes**." We provide some insight into how you can find, identify, and then study organizations who somehow manage to rise above the commonality of everyday life, and do wonderful, innovative things.

- Last, but not least, when you think about innovation, don't get caught up in the current hype that can so often swirl around new ideas. You must always carefully analyze the current round of "cool" business ideas and "**Jump Off!**" when you realize that there could be an innovation band-wagon effect underway.

Ready, Set, Done. • How to Innovate When Faster is the New Fast

Simply Uncomplicated

Rethinking innovation-
what the heck is it, really??

S O IF we're in this world of high-velocity change, and innovation is critical, where do you start? By understanding that innovation is about much more than developing new products to take to market.

Traditionally, most of the discussion and news media around innovation has focused on "cool companies" that bring "cool new products" to the marketplace. For a long time, innovation has seemed to be all about new products, and the design that goes with such new products. Look up a traditional award given to a company for innovation, and you'll find that industrial design esthetics have played a big role in the innovation that has occurred.

While industrial design, and the new product and service development that goes with it is certainly a big part of what makes an organization successful, it is only a small component of what we might consider the "innovation opportunity."

Truly innovative organizations, and the leaders and staff within them, realize that innovation can occur with anything: operations, customer service, business processes, the ability to enter new markets, revenue enhancement opportunities, not to mention corporate and workplace structure, corporate culture and attitudes and just about everything else!

Innovation is about everything an organization does and how it does it. It isn't really that complex to understand – it can really be boiled down to three key goals:

- Run the business better

- Grow the business

- Transform the business.

It becomes quite clear that there is plenty of opportunity to do things differently if you view innovation in these simple terms.

Take the concept of "running the business better." There is plenty of opportunity in every organization for operational innovation; that is, doing what you can to run the business better. This type of innovation involves a continuous effort to change, improve and redefine business processes, whether they involve customer service, HR practices, logistics and shipping methodologies, purchasing processes or just about anything else. There is endless potential for improving the way the organization works, and there are plenty of opportunities for innovative thinking with respect to the way things are done.

Second, make sure you understand the opportunities from "growing the business," or what might also be called "revenue-focused innovation." Don't restrict yourself to thinking that it is only about enhancing revenue through the development of new products or services, think about business model innovation. For example, new business ideas involving expansion into existing markets or new ways of reaching the customer that weren't previously possible (or that no one had thought of before). There might be opportunities for revenue enhancement through partnerships that permit the rapid entry to new markets, or perhaps, revenue focused innovation that comes from evolving a product into a service, thereby avoiding the dreadful path of competing on price.

Innovation is about everything an organization does — and how it does it.

Last but not least, always keep in mind the concept of "transforming the business." Transformational innovation involves taking a look at the way the organization is structured, and thinking how it might be able to work smarter, more efficiently and with better results by changing the skills makeup of the group. Do you need to put in place a structure that fits the unique and ongoing challenges that you are being presented with as the new century unfolds? If not, do you have a flexible structure that helps you adjust your capabilities on an as-needed basis? Do you have an organizational culture that permits rapid transformation of the way you work, such that you can respond quickly to new challenges? If you don't, then you need to be thinking about transformational innovation.

It is entirely possible to pursue an innovation strategy encompassing all three areas of opportunity: running, growing and transforming the business. If you begin thinking about innovation from these three perspectives, you'll come to realize there are extensive opportunities that can come from innovation.

It's quite possible to restructure your operations such that all three innovation goals outlined here are achieved.

I worked with one client within the patio furniture industry. In the early part of the decade, they had seen quite a bit go wrong: the emergence of offshore Chinese competition; the rapid reduction of margins on product; and the loss of several key customers. Their strategy to work their way out of the challenge was to focus on streamlining a lot of existing, inefficient business processes so that staff spent less time on paperwork and were able to take on more valuable roles: that was running the business better.

They also transformed the business by revamping their manufacturing process – they made flexible, short production runs a key part of their capability.

Finally, they ensured that their sales staff were trained to approach the product in a different way, such that they could make sure that existing and new potential customers understood that they could now produce to their unique specifications. They turned their sales energies towards the custom production of product, rather than the same old traditional product lines. That provided for the business growth that they needed to survive the challenges that had impacted their marketplace.

Put all three together, and there was a tremendous amount of innovative thinking going on.

Innovation Oxygen

Innovative, creative organizations do things differently

WHAT IS it that allows some companies to achieve innovation breakthroughs? What do they do differently that allows them to restructure themselves into efficient, collaborative teams that constantly search for the next great opportunity? How do they stay focused on operational agility, and spend barely any time on corporate politics?

There are a variety of reasons: first and foremost, everyone within the organization is firmly looking at a clear sense of goals, direction and purpose. But beyond that, successful innovative organizations share a few traits.

- **People have an open mind on innovation.** As outlined in the previous chapter, people know that in addition to R&D, innovation is also about ideas on how to "run the business better, grow the business and transform the business." They know that innovation isn't some deep, complex process reserved for the elite, they know it is a key component of what they are expected to do on a daily basis.

- **Ideas flow freely throughout the organization.** Hierarchy has disappeared, replaced by a structure that supports the instantaneity of ideas and insight throughout the organization.

- **Subversion is a virtue.** While the organization has to pay mind to the culture of compliance (such as the requirements of Sarbanes Oxley) that has become common through the last few years, it also ensures that stepping outside the rules for issues beyond compliance is not a "bad thing," indeed, breaking the rules within an expected set of ethical norms is expected.

- **Both success and failure are championed.** People aren't held up to ridicule for trying something out. Instead, their attempt is heralded as a breath of innovation oxygen, critical fuel that will fire the organization up for going forward into the future. Only through repeated efforts and failures comes success.

- **Leadership dominates, and managers are minimal.** There are many, many leaders who encourage innovative thinking, rather than managers who run a bureaucracy.

- **Difference dominates.** There are creative champions throughout the organization – people who thrive on thinking about how to do things differently. Rather than being a group of people of common comparisons, there are lots of oddballs about: lots of people who see the world in a different way.

- **A sense of urgency rather than defeatism.** Rather than stating "it can't be done," people ask, "how could we do this?"

- **Responsibility is clear.** The word "innovation" is found in most job descriptions as a primary area of responsibility, and a percentage of annual remuneration is based upon achievement of explicitly defined innovation goals.

Every organization should work to develop innovation as a core virtue – if they don't, they certainly won't survive the rapid rate of change that envelops us today.

People know that innovation isn't some deep, complex process reserved for the elite: they know it is a key component of what they are expected to do on a daily basis.

After a day on the beach with my sons, I had an innovation brainstorm: and wrote a blog entry, "Why Innovation Thrives in the Building of Sandcastles." It struck a chord with a lot of people, even Business-Week highlighted it several weeks later.

Here's why there is an intense amount of innovative thinking that seems to thrive in the building of sandcastles:

- Hierarchy has disappeared. In most cases, sandcastle building does not require a boss or a reporting structure. People just pitch in and do what needs to be done. The lack of a hierarchy is implicit to most successful teams.

- Creativity is implicit. Anyone can build a sandcastle. There are no rules or preconceived notions, other than some sand and water. The same thinking should drive corporate innovation efforts. Make do with what you've got and what you can find, and use creativity as your main asset.

- If it doesn't work the first time, do it again. It's inevitable that a rogue wave will destroy your work. This only encourages you to fix the design, or rebuild it altogether. Setbacks are meaningless, and indeed, are part of the plan.

- Experience doesn't cloud insight. Parents listen to kids, kids get bored and move on to another rampart and do something awesome. The key to sandcastle building is the combined insight of several different generations, likely one of the most important foundations for success in corporate innovation today.

- Everyone picks up on the passion. People just join in and help to build. Eventually beach-neighbors join in, and the growing castle becomes a big collaborative effort. Organizations that can build similar levels of interest in the concept of innovation don't simply succeed – they exceed!

- Feedback is instant. You know right away how well your design works, particularly if it is at the water's edge, since everyone will make a comment on it as they walk by. That parallels the instantaneity of today's markets: things are changing so fast, that you must have a constant ear tuned into what your customers are telling you.

- Competition is easily scooped. Need new ideas? Want to learn from the competition? Spend a few minutes walking up and down the beach and check out the other sandcastles. Study their design, their assumptions, and see how you can improve upon them. Do the same in the corporate world. Develop a finely tuned radar that signals to you how and where your world is changing.

- No idea is too dumb. There's not a lot of criticism and bias in the building of sandcastles. Any idea is welcomed. People can contribute the skills they have. Everyone is a designer, a builder and an owner. Somehow the combination just works.

- The reward is clear. At the end of the day, a great sandcastle provides a sense of accomplishment. Photos are taken, and the team talks about the experience. That's why every innovation effort needs to be celebrated, highlighted, and championed into the corporate record.

- It's risky. As quoted in the poem "Anyway," "What you spend years building, someone could destroy overnight; build anyway."

- It's fun. Enough said. If an organization approaches a problem the same way, innovation and creativity can thrive.

Navigating Innovation

It's the approach that counts

S O INNOVATIVE organizations know what innovation is all about, and they do things differently as a result. What else do they manage to do? They ensure that they have a leadership group, not only at the top of the organization, but throughout the organization, that is focused on the idea of innovation.

You need to do the same thing. To respond to the reality of the world of change, you need to ensure that you and your organization are forward looking and forward thinking. To achieve this it is critical that you get into a leadership frame of mind, even if you have day to day management responsibilities.

Some people might interchangeably use the terms management and leadership, yet they involve distinctly different traits and responsibilities when it comes to the future, change and innovation.

Managers are involved in the day to day direction of an organization. They ensure that procedures are followed, that required actions are taken, and that everything runs as smoothly as possible. Management is a critical skill, one that is not easily attained, and when it is done well, can cause an organization to shine. A great manager is akin to being the captain of the ship, steering the organization on a daily basis through difficult, choppy and often unpredictable waters.

The day-to-day role of managers is to ensure status quo:

- **Maintaining traditions.** They talk of corporate history, and use organizational nostalgia as an excuse for inaction. Old glories and past corporate successes are the focus, rather than the opportunities of the future.

- **Containing risk.** In this era of Sarbanes Oxley driven corporate compliance excess, it has become all too easy for many managers to lay down the law: it's time that everyone in the organization buckle-down, do the right paperwork, and sign and initial every piece of paper! While we certainly live in an era that demands respect and attention towards such matters, an inane focus on risk containment can lead to the death of innovation. Organizations can quickly find that the dedication towards "minimizing risk" has killed the potential for any type of new thinking.

- **Concentrating on day to day details.** Importance is given to time sheets and other structured items. A huge amount of time is spent on minor details, and unimport- ant procedures, rather than time spent thinking about trends, challenges, threats, and other forms of creative thinking that can lead to opportunity.

- **Short term focus.** Everyone is fighting fires, and no one is watching to see if a massive wildfire has broken out. There seems to be an incessant focus on the issues that might impact the organization next week, rather than the trends that might have a more significant impact within the next year.

The day to day role performed by management can stifle innovation.

Leaders, on the other hand, are entirely different. Over the years I have seen many leaders, but they can all be put into one of three categories: those who watch things happen; those who make things happen; and those who sit back,

Leaders prepare their organizations for the future by defining direction, encouraging innovation, and effecting change.

oblivious to what's happening (you can easily identify them, they're the ones who sit back after their world has been transformed and ask, "whoah, where'd that come from?")

You want to be in the second group – those who make things happen. These types of leaders provide an overall sense of direction for the organization and inspire individuals to pursue opportunity. They change the frame of mind and frame of reference for the organization in terms of goals. From this perspective, a leader defines where the boat should go, before the voyage actually begins. They are future oriented, innovative, and adaptable. Innovation requires leadership, because it involves doing things that are new and different.

Innovative leaders focus on:

- **Building a shared vision.** They make sure that everyone is firmly focused on the opportunities and challenges of the future, rather than the battles and problems of the past. Leaders prepare their organizations for the future by defining direction, encouraging innovation, and effecting change.

- **Providing for an open, innovative culture.** They establish an overall organizational culture in which everyone is firmly focused on the future while managing the present, such that threat is turned into opportunity and agility becomes the cornerstone for success.

To accomplish this requires certain key leadership attributes:

- The ability to link the corporate mission of today to the major trends and developments that will influence the organization through the coming years;

- A leadership style that encourages a culture of agility, one that allows for a rapid response to sudden change in products, markets, competitive challenges, and other business, technological and workplace trends;

- The ability to establish and encourage an organization wide "trends-radar," in which all staff keep a keen eye on the developments that will affect the organization in the future;

- A culture of collaboration, in which everyone is prepared to share their insights, observations and recommendations;

- A corporate culture in which staff are encouraged to not only deal with the unique and on-going challenges of today, but are open and responsive to the new challenges yet to come;

- A performance oriented focus in which people are encouraged to turn future challenges into opportunities, rather than viewing change as a threat to be feared.

One of the biggest challenges for leaders to overcome will be to keep the future workforce and organizational partners engaged. This reality provides food for thought: maybe the real path to innovation will come through a strategy we might call "boredom avoidance."

A recent survey indicated that 63% of Gen-Y's – those currently in their 20's – plan on staying less than two years with an employer; 52% of them think it is easy to find a new job; and 49% think that management isn't doing a good job preparing them for leadership and management positions.

Is it any wonder that the average age of a Nintendo player is in the mid-30's? This is a generation that has grownup in a world of fast response, rapid reaction, and instant interaction.

Every generation now entering the workforce is driven by boredom with process, routine, and the "same-old." They're constantly looking for something new, and hence, need to be engaged and involved.

That is going to take a unique form of leadership, and that's where innovation comes into play – the upcoming generations are naturally creative, and are just looking for opportunities to do what they do best. They just might excel at the attributes and attitudes outlined in this section: and might be the most innovative organization that we've ever seen.

Only if they can be prevented from becoming bored out of their skulls!

Step Forward

Key success factors for innovative organizations

WHEN IT comes to leadership, it is important to ensure that your organization is relentlessly focused on the future.

As we mentioned in the previous chapter, there are three types of leaders in the world: those who watch things happen, those who make things happen, and those who sit back, oblivious to what's happening. It's best to be in the second group, those who make things happen.

To those organizations who are focused on the opportunities of tomorrow rather than the challenges of today, innovation comes naturally. By building a culture that is relentlessly tuned into that reality, they easily succeed in doing things differently. In doing so, they're the true innovators – they've been able to throw off the shackles of yesterday's problems in order to concentrate on what they could be doing to adjust to tomorrow.

There are several key elements to this successful, forward oriented innovation culture:

- **Growth orientation.** They've managed to instill a culture that has everyone thinking about what can be done – they are forward oriented. It's a culture in which people are thinking less about the problems that have occurred, and more about the cool strategies that could be pursued. They don't run "change-management workshops": they have strategic sessions on "growing the busi-

There is no more debate about the need to break down silos; they are gone. What remains is a desire to learn from each other, and build on common insight.

ness." It's not an easy task, but innovative organizations have managed to get their people away from "right now" to "our next step."

- **The ability to cost-manage and grow at the same time.** A company that is focused strictly on cost management is likely to be an innovation laggard. Innovative organizations know that cutting cost and operational excellence is only part of the overall innovation equation. They also ensure that at the same time they reduce costs, they are actively working on growing their market, learning how to do things differently, discovering new ideas, and seeking opportunity.

- **A translatable vision.** Every single innovative organization has, at the root, at least one, simple, concrete idea that defines their innovation scope. For example: they might define their future in terms of their ability to utilize external, complex skills. "We know we don't have the depth to generate innovation ideas entirely internally – so we will seek innovation partners to help us drive growth in our sector!" With that type of concrete vision on the issue of skills access clearly defined, the process of doing it becomes that much easier.

- **Time to market is critical.** The organization relentlessly lives and breathes the mantra, "it's all about our ability to get the product/service to market on time." With that key goal, they manage to harness their energy towards a growth and ideas agenda.

- **Internal collaboration.** The organization has gone beyond seminars about teamwork, to a culture in which ideas are instantly shared, debated, transferred, analyzed and recomposed. There is no more debate about the need to break down silos; they are gone. What remains is a desire to learn from each other, and build on common insight.

- **Transition at the top from managers to leaders.** Innovative organizations have transitioned the roles of management. They've moved away from such mind-numbing activities as day-to-day oversight, implementing procedures, generating reports, and all the other routines. Instead, they're redefining the organization by pushing decision-making to the lowest capable level; they define goals and outcomes that staff can pursue; they are putting in place collaborative and market oriented feedback loops. In other words, they're thinking about all the things the organization should be doing to survive and thrive in a fast-paced market.

- **At every level, there is a tactical to strategic transition.** At the same time that the role of management has evolved, so too has the role of staff. As part of the shift to innovation, the organization has transitioned the roles of many staff so that they provide a more strategic role ("what do we need to do to meet this new challenge?") rather than routine tactical activities ("I need to get this reconciliation signed off!") They've done this by automating a lot of the routine, day to day processes that can choke off innovation. They're freeing staff up from the performance of the mundane, to the enhancement of opportunity.

- **A partnership orientation.** Innovative organizations recognize that they can't do it all. They look to actively work with their complexity partners in order to be able to do innovative things at the pace that change demands.

Innovative leaders are redefining the organization by pushing decision-making to the lowest capable level; they define goals and outcomes that staff can pursue; they put in place collaborative and market oriented feedback loops.

Partnership is a key word: "we might not be able to generate all the ideas, but we sure as heck are going to know how to find them and capitalize upon them."

• **Global skills access is a key success factor.** Big or small, you can only drive innovation if you can access the complex talent needed to take you forward. Innovative companies have mastered the article of just-in-time skills access; they can access and focus a unique set of skills for a unique purpose at a moments notice.

Remember, when it comes to innovation, it is all about leadership. Only those who make things happen can provide for a focus on real innovation.

Undertaking a tactical to strategic transition with your staff is quite possibly one of the most innovative things that you can do – and yet, you might find your staff has some deep suspicions with respect to your efforts.

Innovation often involves the re-organization and re-deployment of staff, and because of that, some of your staff might interpret signs of change as an indication that: "people are going to be let go." Yet that is often not the goal of a tactical to strategic oriented innovation effort: the goal is to ensure that staff move away from anything that is routine, day to day, or task oriented, into a role in which they apply their skills and talents in a way that helps the organization change, innovate, grow and prosper.

A client within the kitchen cabinet industry had a tremendous number of routine, manual business procedures in place. There were frequent errors in orders, all of which were quite costly due to the need to rework various components; tremendous inefficiency as a result of many paper-oriented procedures; and lack of real market insight, since the sales team really had no idea of the track record with any particular client.

After a deep, expensive project to bring their information systems up to the latest standards, they saw real transformation: they were able to take 3 1/2 people off of the 'order desk' and put them into a role where they were involved in the sales and support of custom kitchen cabinets. Within a year, they saw that line of business grow to 40% of their sales and 60% of their profits.

Pinpoint Analysis

How can you identify areas/opportunities for innovation

NOW WE'RE getting somewhere – by now, you've got a good idea of the different elements that make up for successful innovation. What's next?

Start out with the mindset that there's a lot of opportunity to do things differently, if you think about innovation as the "fixing of problems." As has been indicated to this point, innovation need not be strictly the development of new products or services; there's plenty of opportunity to be had by focusing on existing business methods, processes, structure and methodologies.

One of the easiest ways to find opportunities for innovation is by looking for the things that are "broken," and figuring out how to fix them! For example:

- **Look for your blind spots.** Where are you lacking information which leads to missed opportunities in your marketplace, failure to provide stellar customer service or excessive operating costs? Undertake a simple "information inventory," with the idea of identifying where you are lacking in the critical insight you need to be innovative.

- **Identify where you have been making decisions while in the dark.** As part of inventory, identify the situations where you've missed opportunities, have made the wrong decision, or have been misfocused because you simply had the wrong information, or have had the right information at the wrong time. You could call this your insight deficit. That's one of the first areas that you can focus on for innovative opportunities.

- **Highlight your big failures.** Where did things really go wrong: where are there excessive issues concerning quality, customer relationships, time to market or other aspects of your organization where you have simply failed to do what should have been done? If you take the time to find and confront your biggest foul-ups, you'll discover plenty of opportunities for innovation.

- **Identify the biggest threats.** Undertake an assessment of the biggest challenges that you might face on a six month, one year and five year horizon. Where might new competitors emerge? What current shortcomings might cause you to miss opportunities? Where might "blurring" within your industry or market sector occur as a result of continuous product or service evolution?

- **Focus on fear. Get scared.** Look to your competition for insight on what you might be missing. What is the most innovative thing that they are doing? What levels of risk and creativity are they willing to deal with that you aren't? And what is the real impact if you don't do something drastic right now to catch up?

- **Look for the "lost causes."** Where are you simply wasting a lot of time, doing things that you shouldn't be doing? For example, do you have a sales force that spends more time looking for information and insight into specific customers than they do selling? A marketing team that is still focused on fighting the brand wars of yesterday as opposed to the ridiculously fast changing consumer preferences of today? Where are you simply spending too much time undertaking futile activities?

- **Identify resource-leakage.** Where is valuable talent being unnecessarily diverted? Is your management team constantly fighting fires, rather than focusing on strategy and opportunity? Do you have a customer service staff that is busy putting data into twelve different databases, rather than fixing customer problems? Take the time to figure out where you are wasting a lot of valuable time doing things you shouldn't be doing.

- **Pinpoint your weaknesses.** List the things you can't do, but that you should be able to do. Where can you see shortcomings in your capabilities compared to the competition or high fliers in other industries? Who are you losing business and employees to and why? And what could you be doing differently to avoid these problems?

- **Find your routines, and break them.** What are you doing that is absolutely, positively ridiculous – but you keep on doing it because you've always done it that way?

Innovative organizations are full of people who have been through the "tactical to strategic transition." That's one of the reasons why ideas flow freely: people are firmly focused on the opportunities of tomorrow rather than the problems of today.

How can you achieve this? By changing the focus of your staff, so they think differently. You want to ensure that they move away from anything that is routine, day to day, task oriented, into a role in which they apply their skills and talents in a way that helps the organization change, innovate, grow and prosper.

How can you hope to confront the future and develop a culture of innovation, if you are unwilling to talk about the big challenges that you are faced with?

That's a mystery that I still think about when it comes to one particular client, a government agency. They had engaged me to come in and provide a leadership session that would help them be more responsive to the new demands being placed upon them.

In the planning stage, we spent a great deal of time identifying the high-velocity changes that were impacting them, and many of the concrete, practical steps that they might undertake to deal with these challenges.

It looked like it was going to be a great session: we would have a full day, with a workshop, examining future trends and innovative ideas from a wide variety of perspectives.

Yet just a few days before, they contacted me, and decided that it was not a good time to "move forward." Their organizational challenges were too deep; they didn't think that staff and managers should be talking about the necessity to move suddenly and rapidly to deal with those challenges; they didn't think that we should "rock the boat."

In other words, they chickened out.

At the same time, the rate of change continues to increase; the demands and expectations of the service they must deliver continue to go up; the need for change has not gone away.

Rebels Yell!

Offbeat sources for ideas

I F FORWARD-ORIENTED insight is critical to innovation, how can you figure out which trends are going to impact you in the years to come? One way is by seeking insight from beyond your usual information sources.

Often, you'll read the same newspapers, chat with the same office staff, go to the same conferences, network with the same industry people, and read the same professional journals. By doing this, you are ensuring that you regularly get the same old hum-drum insight that probably shields you from seeing what is really going on.

Ask yourself this: is your view of the future inside a bubble – if yes, then you've got to burst the bubble and get out to see what is really going on and what might really impact you!

When you are in the same-old information bubble, it is all too easy to become isolated and focused on the issues of the day. That's how you end up spending so much time on management, and precious little time on leadership, the role that provides opportunities for innovation.

One of the most overused phrases is that you've got to "think outside the box." Go beyond that by ensuring that on a regular basis, you 'step' right out of the box. (I joked at one conference that they should throw the box out of the window – and jump right after it.) One of the best ways to pick up an innovative idea is to work hard at finding new sources of inspirational, offbeat information. There are a few sources you should think about.

Listen beyond the grassroots.
You can't listen only to the "usual suspects" to spot the trends that will affect you – you have to go beyond your roots to listen to what others are saying as well. 21st century leaders recognize that everything in their industry is being affected by events, trends and developments far beyond the traditional borders of their own industry.

One of the best ways to build your creative insight is by taking the time to place yourself in completely different circumstances on a regular basis.

For example, pick 2 or 3 other conferences each year – in completely unrelated, different industries or professional areas far beyond your skills base or the industry you work within. Go and listen – and see what another industry is saying! You might discover some wonderful ideas that could be used in your own industry.

You can also subscribe to professional or trade publications for other industries. Pick a few unrelated industries at random – and sign up for their magazines, publications or newsletters.

You might be surprised how invigorating an experience it can be to open up your mind to what is going on elsewhere. You will find that it helps you to discover the trends that will affect you in the future, long before your traditional trends radar might have picked them up.

Listen to those who you don't like.
Often, the trends that will affect your industry can be found in the offbeat chatter by those who are busy redeveloping the future because they have completely different ideas about the future.

Seek out the rebels in the industry — you might not like what they have to say, but often, they are right in what they will tell you.

These leading edge trendsetters – lets call them rebels – are often at odds with the typical assumptions that under lie an industry. They're the rebels in the crowd, eager to cast off the past in order to develop a future that will be very different. They're busy tearing apart the conventional business models that have guided your industry; they have different ideas as to the nature of the product or service that is delivered; they are all too eager to change everything around them in order to create the future as they see fit. They are often marginalized, simply because their aggressive attitude in changing the future can make them rather unlikable by many.

What should you do? Learn to learn from them! Great leaders recognize that while there are lots of people who have an attitude, outlook, culture and approach to life and business that is completely at odds with their own perspective, they are willing to listen to what others have to say, because change often emanates from such people.

Listen to what the exodus says.
There are quite a few people in our economy today who don't work within the traditional corporate model anymore. They often have very different ideas about the future than those who still work within the traditional business model. And often, they've chosen to bail out of structure in order to prove their ideas.

For example, many young workers continue to reject the traditional career path of long term careers with large organizations, instead establishing themselves in small, micro-organizations that can provide needed skills to a corporate audience regardless of where they might be. They are a wonderful source of insight for innovative ideas.

Then there are nomadic workers – those workers who were laid off in the last 10 years through a variety of recessions – and who have established small, home-based businesses from which they provide their skills to a global audience. They're working within your industry, but often view the world in a very different way. What can you learn from them that might provide areas for innovative opportunity?

Get young.
Throughout the next year, take the time to listen to young people – anyone ten years or more younger than yourself. They're building the future right now, and you'd do well to understand it. Their future is hyper-active, interactive and multi-tasking – this generation gets bored quickly, and they are entering your workplace. They are also becoming your new competitors. Don't expect them to subscribe to the same old beliefs as to structure and rules, working hours, and corporate culture, or business models. You won't survive in their future if you don't take the time to understand what they are doing, talking about, and thinking.

Is any idea too offbeat? too farfetched? Not in my books!

Consider the concept of a "printer of things." Science is now evolving quickly enough that some researchers foresee a day, not too far away, when our homes and offices will have small digital fabricators (also known as "fabbers"), electronic devices that will assemble almost anything.

Need a new dishwasher? Download one, and it will self-assemble itself in your home? Seems like a ridiculous idea?

Maybe so – yet the concept has been around for a long time. It was first proposed by John von Neumann, a leading scientist involved in the development of atomic weapons and other technologies, and was often referred to in Star Trek shows as a "replicator." Although it sounds terribly too futuristic, so too was the idea of home computers more than 30 years ago. Do a search online for the phrase "fabbers," and then think about whether this is really a whacky, off-beat idea, or the sign of something yet to come.

Ready, Set, Done. • How to Innovate When Faster is the New Fast

Keep on truckin'

A unique source of innovation insight

INNOVATION CAN come from offbeat ideas, rebels, and those who think differently. Ideas for innovation can also come from those you might least expect would be busy with new ways of thinking.

So here's a point to ponder: some of the most innovative organizations and individuals happen to work within the trucking industry, simply because they seem to best understand the potential for real innovation.

Truckers? Did an image just spring to mind?

If so, drop that image, because the trucking industry has been at the forefront of what we might call the "logistics innovation revolution" involving:

- operational efficiencies,

- revenue growth and

- organizational transformation

all in one swoop.

It is those three elements that often are the essence of innovation for any organization – "running the business better, growing the business, and transforming the business."

Layering a service element over top of your basic business model could transform and build your business.

Consider what the trucking industry has been doing. Over the last ten years, a wide variety of trucking and shipping organizations – ranging in size from mammoth organizations such as FedEx and UPS to the smallest of local carriers – have been at the forefront of an effort to reduce their costs, grow their revenue and transform their line of business. They've been doing this with the overall strategic objective of assisting their customers in the transformation of their own business operations.

Collectively, they grabbed on to a simple premise: trucking organizations could increase the value of their product and service if they could help their customers run their business better. They could transform their own operations and those of their customers by taking on the role of business partner to their clientele. One of the ways they have done this is by layering a service element on top of their basic business model.

Look at what trucking organizations offer today: simple tools that allow their customers to build efficiency into the process of arranging for and tracking shipments. These tools allow the trucking organizations to reduce their own cost of doing business, which is the idea of "running the business better," or operational transformation.

Sophisticated planning tools help them maximize the load on their departing trucks. Many organizations have also developed systems to help them schedule last minute or partial loads for immediately departing trucks, which they offer to clients at a reduced rate. Both of these strategies, along with other innovations, have helped them to grow their revenue.

Last but not least, they determined there was an opportunity to transform the very nature of what they do: they could take over the entire shipping operations of many of their client organizations, which led them into the "logistics" business. In effect, they've transformed themselves into an effective partner with their clients, and have achieved transformational innovation.

Through each of these three key strategies, they've mastered the nuances of leading edge innovation, even while many folks consider them to be "just a bunch of truckers."

We'd do well to learn from their brilliance!

Here's an interesting way to think about innovation: if you do something to solve an internal problem, and you do it really, really well you might find that it can also solve the problems of other people – and so your internal solution becomes a new service or product that you sell!

One company I spent time with was involved in manufacturing 16-foot and longer wood mouldings. They saw a unique, innovative opportunity when they took a new look at "what they did well."

The organization had placed significant effort into developing a sophisticated logistics (shipping) application, so that it could better manage its product shipments, and coordinate those shipments with its primary customers (large retailers and distributors).

They realized that they were one of the few companies out there who had developed specialized expertise around the shipping of 16+ foot long packages – and hence opened a new "third party logistics" business, in which it began to provide this service to other organizations who had need for such unique shipping capabilities!

They started out by figuring out how to do something to run their own business better, and in the process, discovered they had also developed something to grow their business. That process caused them to rethink how they approached issues of business strategy, which of course leads to transformation of the business!

The Word is: Transformation

Are you watching the major transformations or just the piddly stuff?

P ART OF the art in discovering opportunities for innovation is by learning how to spot the trends that really matter.

Not the small, incidental, day-to-day, one-to-two-to-five year trends. Instead, you should be looking for the "big transformations" – the sweeping, massive, significant types of change that causes everyone to sit back twenty years later and ask, "Wow! Where did that come from?"

Consider, for example, the world of health care and life sciences. Certainly everyone is aware that current trends indicate that the challenges are vast and the opportunities are significant. There's a looming shortage of skilled workers, dramatic rates of discovery of new knowledge, the rapid emergence of new medical methodologies, and other forms of significant change.

Yet, with everyone focused on these issues, most people are missing the big, long range transformation that is underway: we are in the midst of a fundamental and significant shift in healthcare philosophy and medical research that makes every other trend within this industry pale in comparison. We're rapidly moving from a world in which we "react" to disease and illness after it has happened, to one in which we will be doing far more in advance to "prevent" specific health care problems from occurring in the first place.

When looking at the big picture, you should always step back and look for an even bigger picture.

The driver for this massive change is the emergence of extremely specialized and highly personalized medical treatments based upon your own particular DNA.

Preventative medicine has already become a part of the health care system: for example, a simple pap smear test has resulted in a 70% reduction in cervical cancer since the 1940's. Yet, it is estimated that clinical diagnostic spending of this type makes up only 1% of global health expenditures.

DNA "sequencing" is set to change that, as it allows researchers to examine an individual's DNA, and determine their risk for developing particular diseases or medical problems. Already, a test has been developed that examines a few hundred strands of DNA, from which a prediction can be made of your risk of developing cystic fibrosis (CF). The test accurately identifies the unique DNA strand in 88% of Caucasian CF carriers and 69% of African Americans CF carriers. Expect the degree of accuracy to only continue to improve in coming years.

Consider another example of a "deep transformation change" in another critical industry: energy supplies. In the last few years, we have seen lots of price volatility as a result of the unexpected, such as Hurricane Katrina. One result of volatility has been a renewed interest in bio-fuels, and in particular, the opportunity that exists from the creation of ethanol from corn and other crops. Ethanol has grown quickly to become a rather significant industry.

Yet the big transformation that we will see is the rapid changeover from "first stage" ethanol companies to "second stage" bio-refineries. The first group consists primarily of agricultural companies, using their insight into the science of agriculture, to develop production systems that convert grain to ethanol. The second group are the big oil companies, who are bringing to the industry their insight into how to build big, production oriented, cracking and distillery methodologies to the world of ethanol. In doing so, they will be transforming a high velocity industry into a faster and more complex industry involving "bio-refineries."

This transition is both massive and sweeping in scope. Royal Dutch Shell Europe, for example, is constructing a "second generation" plant that involves a capital investment of $2,000 per ton to construct, compared to $190 per ton for a first generation bio-fuel plant. The switch to bio-refineries will be significant. Some estimates suggest that the major oil companies will be able to grab upwards of 17% of the global biofuel market within a few short years.

It's by watching for and identifying such massive shifts – a switch from reactive to preventative medicine, or the emergence of a bio-refinery industry– that you can spot real areas for innovation and creativity. That's why when looking at the big picture, you should always step back and look for an even bigger picture.

In the "big transformations," you can witness the birth of entirely new careers as well as new industries, and in doing so, find countless opportunities for innovative thinking.

In the world of health care, for example, we are seeing the emergence of a new class of health care professional, the "hospitalist." These individuals, usually with a medical background, provide a simple role to patients: they steer them through the ever increasing complexity of the medical system. That's no small challenge!

Why do hospitalists exist? Because we are now witnessing a furious pace of change with the advancement of medical science. Medline, the online research service, suggests that if a medical professional in 2004 wanted to keep up to date with advances in adult coronary heart-disease, they would have had to read 3,672 articles on the subject. (That was in 2004, how many more would they have to read today!!)

If they spent 15 minutes per article, it would take them 115 -eight hour days to read them all. Considering that there are about 12,000 known diseases, there's a staggering amount of new medical knowledge out there.

And that's a lot of knowledge to keep on top of: with the results that an entire new profession has emerged: people who have the focus of helping to navigate patients through the ever growing complexity of the medical care system, largely driven by ever more rapid growth of new medical knowledge!

Innovation Heroes!

Find the most innovative industries and use them as a source of insight

OFTEN, ONE of the best ways to discover ideas for doing things differently can come from studying other organizations or industries. There are many organizations out there who aren't innovative and are stuck in a rut; there are others who are extremely innovative, at the same time that there are a lot of laggards. It does not matter if you are a large multinational or a small company, you can still be stuck in an innovation rut.

One of the best ways to discover new and creative innovation ideas is by studying those who are moving forward at a really fast pace. They might be within your own industry; quite often, they will be in a completely different industry.

Organizations or industries that are subject to extremely high velocity are often the most innovative. They are busy working with the challenges that exist, and are being as creative as possible to deal with those challenges in order to turn them into opportunity.

Regardless of who and where they are, they share several things in common: they're busy experimenting, adapting, evolving and changing. They're working hard to make sure that the essential concepts of high-velocity innovation – run the business better, grow the business, transform the business – have become an essential part of their lifeblood.

If you can spot these organizations, you can learn from them, and become inspired by them. They can be a wonderful source of creative ideas!

So how do you find them? By looking for the telltale signs of companies or industries who are faced with all the challenges that the high-velocity economy can throw at them. Given the challenges, the organization or industry will tend to have people who are more innovative, realistic, practical, and open to new ways of thinking. They are likely to be more forward oriented and creative. They will be working to rapidly adapt to changing circumstances, and will be collectively seeking complex solutions to unique problems.

Several signs can provide you with insight as to whether the company is dealing with extreme velocity and is therefore a real innovator. Look for these characteristics:

- **They are significantly impacted by faster science.** The fundamentals of the science within the industry are evolving at a furious pace as a result of the infinite idea loop. It is evident that the discovery of new knowledge within the industry is occurring at a faster pace than within other industries.

- **More competition.** Business models are changing quickly, with a lot of new competition appearing on the scenes, as the industry begins to blur and change.

- **A faster degree of product/service innovation.** The industry is widely known for being innovative, with a constant stream of new products or services coming to market.

- **More operational innovation.** There is a lot of fundamental change within the industry in terms of business models, marketing methodologies, customer relationships and other unique changes.

Organizations or industries that are subject to extremely high velocity — that is, significant amounts of fundamental change occurring at a rapid pace — are often the most innovative.

- **Shorter product lifecycles.** Products are coming to market faster than previously, or faster than within other industries, due to the previous four trends.

- **Rising tides require fast change.** Customer expectations are changing quickly in terms of the products or services being offered, because of the furious rates of innovation that are occurring. In addition, there's heightened customer service due to hyper-competition; people know that they must absolutely excel in service levels.

- **A significant creativity capability.** The organization or industry is dominated by creative thinkers; a workforce and management team that is fully focused on doing things differently, in order to respond to the reality of change that engulfs them. Those who kill ideas aren't the dominant force; those who suggest how things could be done differently are at the forefront of action within the organization.

- **A partnership orientation.** The organization or industry is constantly seeking outside expertise in order to help it go forward; it is willing to make use of complexity partners, nomadic workers, skills banks and other partners in order to grab on to ever more important change capabilities. They know they can't do it all, and so they are willing to do what it takes to get access to what they need to get it done.

- **They're plugged in.** The organization or industry is linked into and is feeding off of the ideas from within the infinite idea loop. They are constantly scanning and sifting through the constantly evolving collective insight of the global discussion that is taking place; they are always eager to spot how innovation is occurring outside of their organization, and are busy interpreting what is being said in order that they can use this insight for their own purposes.

Organizations or industries that are subject to extremely high velocity – that is, significant amounts of fundamental change occurring at a rapid pace – are often the most innovative. They are busy working with the challenges that exist, and are being as creative as possible to deal with those challenges in order to turn them into opportunity.

You want to find these organizations, study them, and learn from them – since that will be one of the best ways to create your own innovation oxygen.

A huge amount of my time is spent providing insight to some of the largest organizations in the world on the trends that will impact them.

I'm helping them look at what they must do to adapt to the ever rapid economic change, collapsing product lifecycles, fiercer market competition, the rapid emergence of new competitors, challenging new workforce attitudes, not to mention the necessity of gaining access to ever more specialized skill sets. In doing so, I've come to learn that many leading thinkers of our age truly don't appreciate just how quickly the world is changing : they don't understand velocity!

For example, I often tell the story on stage of a hypothetical "Google-Car" – you order it online, it arrives by FedEx, and comes with a party in a box (so you can invite your friends over to celebrate your new car). I explain to the audience that we live in an era in which Google could choose to become a car company – and could jump into the business pretty quickly if it wanted to. All it would have to do is line up the proper partners for the project.

Although this started out as a hypothetical story that I would joke about on stage it quickly became real when I discovered that Google's founders are now significant equity participants in a new California car company, Tesla Motors. This company has brought to market, rather quickly, a fascinating new electric vehicle. It did so by bringing together a wide number of partners to the project, each bearing their own unique expertise and skills. Its objective – by pass the traditional automobile dealer network by dealing directly with the customer.

In the global economy of today, the capabilities needed to design, build and deliver a sophisticated new automobile can become accessible at the drop of a hat. Well, perhaps not that easily, but learning how to manage a project of such scope and scale will become one of the critical success factors for any organization in the future.

This example just highlights the fact that we live in a time in which things are happening so fast that predictions go from fantasy to reality in but a matter of months.

Today's economic winners excel by putting together rapid, global, sophisticated, knowledge-deep partners. You might need to learn how to assemble such project partners faster than you might think!

Jump Off!

Why bandwagon innovation doesn't work

ONE LAST thing to realize about innovation before we move on to some practical steps that you can take to move forward: you've got to be careful of pursuing an innovation agenda that is driven by hype.

The business world often seems to be driven by a bubble of hype. Every once in a while, some existing phenomena is captured and labeled and trumpeted by the media as being the "next big thing." Many companies jump on the 'next big thing' bandwagon believing they are participating in something revolutionary.

History is littered with the wreckage from those who think that innovation is best found by simply piling on to the latest trend. Their hearts might be in the right place but companies that jump on board this type of bandwagon should realize that they are ultimately destroying any innovative spirit that might have existed within the organization.

"Bandwagon innovation" or piling on to a trend because everyone else is piling on is doomed to fail, for many reasons.

- **It's lazy.** True innovation takes hard work. It involves massive cultural, organizational, structural change. It involves an organization and leadership team that is willing to try all kinds of radical and new ideas to deal with rapid change. An innovative organization can't innovate simply by jumping on a trend. Trying to do so is just trying to find an easy solution to deep, complex problems.

Jumping on a bandwagon can often destroy any innovation momentum that you might have. After the bandwagon effect ultimately fails, people end up feeling burned out, cynical, de-motivated – they'll be prepared to do little when the "next big thing" comes along.

- **It involves little new creativity.** By linking a new approach to doing things with a commonly used phrase (i.e. "social networking") means that people end up shutting their brains down. Creativity is immediately doomed through commonality.

- **It's just a bandaid.** Bandwagon based innovation causes people to look for instant solutions and a quick fix, rather than trying to really figure out how to do something differently.

- **It's misfocused.** It involves putting in a solution without identifying a problem. It's backward in terms of approach!

- **It encourages mediocrity.** It reduces innovation to an "idea of the week," and does nothing to encourage people to really look at their world in a different way.

- **It reduces innovation to sloganeering.** Truly creative people within organizations are tired of slogan-based management. They've seen far too many "radical right turns" and "new beginnings." When they realize that their management team has jumped onto the latest trend bandwagon, their faith and motivation goes out the window.

There will always be a 'next big thing' and as everyone begins to pursue the same idea, a dulling commonality will result. There is little new thinking and little new innovation. This doesn't mean there aren't a lot of opportunities in seeking ideas for innovation from what you see elsewhere. But ideas are just that, and they should be used as a springboard for your fresh, new thinking, rather than something that you can simply copy and duplicate.

Sometimes organizations are so adept with innovation that they don't even jump on a bandwagon – they miss them altogether!

"Innovating after the boat has left the dock!" That's a phrase I use to describe companies or industries who are so blind to the future that they miss the most significant trends that will impact them.

I once spoke to a credit union association and worked to turn their attention to the biggest challenge they would face in the future. To do so, I concentrated on the unique attitudes that GenConnect (those after Gen-Y) and Gen-Yers display towards banking.

First off, I noted that GenConnect are already a financially sophisticated generation: one study suggested that the average annual income for 6 – 14 year olds is well over $1,000. Secondly, the generation before them – Gen-Y – were members of credit unions: some 26% of all credit union members were from Gen-Yers – but, here's the kicker, many of them aren't active members, but simply have an account because their parents had created one for them. Even fewer Gen-Connects had accounts, because fewer of their parents were members of credit unions.

I went on to note that GenConnect is set to become a big part of the future potential financial marketplace: they are set to inherit some $17.8 trillion from their parents through the next twenty years. Already, 1 in 3 have a credit card, but in most cases, with a bank – not a credit union.

How many credit unions were actively targeting this market? 1 in 8.

Activity

Ready, Set, Done. • How to Innovate When Faster is the New Fast

Activity

What you should start doing now to elevate the importance of your innovation efforts

SO WHAT do you do now? You recognize that innovation is critical in the high velocity economy, and that you'll need a different type of organization with a lot of unique, specialized skills to deal with rapid change. How can you take the concepts that you've learned so far and turn them into actionable steps?

That's what we outline in this section:

- Your investment in your innovation efforts will be critical to your success. In the high velocity economy, it isn't just financial capital that will be necessary, there's a wealth of other assets that you can deploy, as outlined in "**The New Money.**"

- You also need to be prepared to think differently as to what to do in this new world of heightened competition (often referred to as the new "flat world"), and an era in which everyone seems to be competing on price. In "**Make Waves**," we outline some of the things that you might consider doing after "your world has gone flat."

- You should also undertake a self-assessment of your organization, to determine if you have the right "state of mind" to pursue a real innovation agenda. In "**Checklist!,**" we outline some of the attitudes which tend to sink innovation efforts even before they begin.

- Innovation causes all kinds of changes and stress which some people can't tolerate. In "**Seeking Support**," we outline some of the best practices that are followed by organizations who have succeeded with their innovation efforts.

- In "**Elasticity!**," we revisit the issue of corporate agility. You need to make sure that you design an organizational culture and structure, as well as business, product, and other plans, that provide for the ultimate in future flexibility.

- In "**Just Do It!**," we provide some straightforward, simple guidance, on what you should do to increase the overall importance of innovation within your organization.

- Finally, you might think about the need for a new degree that can help you with innovation in the high-velocity economy: a Masters of Business Imagination. In "**Degrees of Innovation**," we take a look at the concept of this new (and at the time of publication, non-existent) university-level degree.

The New Money

21st century capital

I N THE 20th century, financial capital was the resource of choice. In the 21st century, given the high velocity economy and the need for agility, organizations will need many different, nontraditional types of capital:

- **Experiential Capital.** In a world in which Apple can toss out the $1/2 billion iPod Mini market overnight in order to enter the new iPod Nano market – it's critically important that an organization constantly enhance the skill, capabilities and insight of their people. They do this by constantly working on projects that might have an uncertain return and payback – but which will provide in-depth experience and insight into change. It's by understanding change that opportunity is defined, and that's what experiential capital happens to be. In the future, it will be one of the most important assets you can possess.

- **Skills Accessibility Capital.** Talent, not money, will be the new corporate battlefront. Simply put, there is so much happening that no one person or organization can know everything there is to know. With ongoing rapid knowledge growth, instant market change, fast-paced scientific discovery and constant skills evolution, getting the right people at the right time for the right purpose will be the key to successful change.

- **Creativity Capital.** It is the ability to see the world differently, and the skill to imagine how to do things differently,

that will be more important than any other skill. This will
bring the needed forward oriented depth that organizations require. When product lifecycles are disappearing,
and market longevity is measured in weeks and months,
the ability to think, adapt, and imagine will be the foundation to provide for necessary change.

- **Generational Capital.** We are set to see the emergence
of the most unique workforce in history, with the wid-
est age-span ever. Boomers won't retire, and kids won't
want to get hired. The result will be a workforce that is
transient, temporary, shifting and flexible. It will be those
organizations who can match up the experience and
wisdom of the aging baby boomers with the insight,
enthusiasm and change-adept younger generation who
will find the most powerful force to be found in business
– an organization that is fuelled by the pure energy of
change-oxygen.

- **Collaborative Capital.** Forget the idea of having a
strategic planning department, and think collaborative
culture instead. Take a look around you, and ask yourself,
who is succeeding today? It is those organizations who
are plugged into the infinite idea loop that surrounds us.
They've dropped any pretense that they can create the
future, and instead realize that the future is being developed by everyone all around them. They have come to
learn that their role isn't to plan for that future, but simply
to listen to it, plug into it, and plug their growth-engine
into it.

- **Complexity Partnership Capital.** In the 20th century,
organizations focused on hiring the skills that they
needed to get the job done. You simply can't do that
today – skills are too fragmented and too specialized.
That's why successful organizations have mastered the
art of complexity supply and demand. They provide their
own unique complex skills to those of their partners who
need such skills. And when they are short on other skills,
they tap into the skills bank of their partners. By selling

Take a look around you, and ask yourself, **who is succeeding today?** It is those organizations who are **plugged into the infinite idea loop** that surrounds us.

and buying skills with a broad partnership base, they've managed to become complexity partners – organizations that spend most of their time focusing on their core mission, and spend less time worrying about how they are going to do what they need to do.

- **Innovation Capital.** Companies that understand that all future innovation comes from the ability to tap into the global innovation loop will thrive; those that follow traditional innovation models, self-centered and insular, will find that their creativity and uniqueness has been smothered.

Enhancing your real world balance sheet with these new elements of capital will prove to increase your innovation capability to a great degree.

In a fast future, you've got to learn and relearn.

Corporate equity isn't just about money, it's the cumulative experience and knowledge of the team.

One of my clients is a major telecom company in the US. They are one of the few companies in the US to take fiber optic services directly to homes, and it takes a lot of abuse from Wall Street analysts for such a big bet.

But here's what might really count: the CEO has stated that the cost of installing fiber dropped 30% in 2005, and that there was a further reduction of 15–20% in 2006.

By the end of 2006, they expect it to cost ½ that of 2005.

That's experiential capital, and that's an invaluable asset.

Make Waves

Innovative, creative organizations do things differently...

WHEN IT comes to innovation, quite a few people don't know how to innovate because they don't know what to do next. Why? They've spent all their time cost-cutting, and have forgotten about all the other stuff they could be doing!

Certainly managing cost has been an important issue, given the emergence of low cost competition from China. Yet, it also begs the question – what do you do after the world is flat?

If your world has become flat, and you don't know what to do next, then put a ripple in it! Change it! Do something different. It's not just cost that is important. That's what innovation is all about – taking inertia and turning it into velocity.

How do you put a ripple in your flat world?

- **Focus on the brand.** In the low-cost economy, brand names still matter if you keep the brand up-to-date and cutting edge. If you do the right things at the right time at the right velocity, your customers will see the innovation in your brand, with the result that your brand will still matter to them.

- **Get religious on "time-to-market."** Hyper-innovation is real – market velocity in every sector is huge as new products are invented and existing products are reinvented. To stay alive, you can't just pump out low cost junk – you have to get the right stuff to the right market at the right time!

• **Go deep with market knowledge.** Every market is being devoured by furious innovation. Ask yourself this: can your sales, marketing and other staff keep up? Maybe not. A fascinating survey in the Birmingham Post, in an article about car dealerships, noted that "…35% of sales staff had little confidence in their own ability to demonstrate hi-tech in-car equipment such as BlueTooth devices and voice control systems." In other words, companies are selling into rapidly changing markets, and yet their sales staff doesn't have the insight of understanding what it is they are selling. That's not good.

• **Increase value.** Think about the rapidity of change occurring in the world of sporting goods. Take a look at a baseball bat: while you might see a piece of wood, it's likely to be a far different type of thing by 2015. It will be highly interactive, enabled with intelligence, and will offer the kid of tomorrow some interactive information on their swing training. The company that does that – and links itself to the heightened expectations of the consumer of tomorrow – will have established some type of new value into a traditional, low value, low cost commodity item. Now that's cool – and that's innovation.

• **Focus on agility.** Forget planning – just do. No one can presume that markets, products, customers and assumptions will remain static: everything is changing instantly. Business strategies and activities must increasingly become short term oriented while fulfilling a long term mission.

• **Seek partners.** We are entering a world that increasingly involves complexity partnerships. Simply put, in the high-velocity economy, you can't do everything at the pace demanded of you: you can only do it if you seek out those individuals and companies who possess a unique skill, suitable at this moment for a specific purpose.

Take a look at a baseball bat: while you might see a piece of wood, it's likely to be a far different type of thing by 2015.

- **Go upside down.** Turn product and service innovation upside down. Look to your customers, suppliers, and just about everyone else for ideas on how to reinvent your products. You just might find that they've already redefined your product, and you weren't even aware of it.

- **Stay motivated.** Folks who have "gone flat" or who "get flat" seem quite dispirited. They have been relentlessly focused on cost, yet there is so much more to the future than becoming a low cost operator. That's what innovation is all about: doing much more than simply "surviving" in a world that has gone flat. You want to put the ripple back into the world through innovation.

Going beyond flat is probably the first step in adjusting the realities of your structure and innovation culture for the future.

One of the most successful methods of avoiding the "flat world" comes through adding a service element to a product. One of my clients did this by providing remote diagnostics and management of their product.

The hyperconnected world of today permitted the embedding of a wide variety of sensors throughout the product. This allowed the organization to actively monitor the product and determine when it might be close to a breakdown or malfunction, thereby heading off any potential problems in advance.

The next step? They stopped selling their customers the product – instead, they sold them a service that consisted of the product PLUS guaranteed service levels, at a much higher price, obviously. The customer is happy with the deal, since they have certainty as to the performance of the product, which could cause expensive delays and excess costs during an unexpected breakdown.

Now that's a ripple!

Checklist!

Is it time for an innovation audit?

MAYBE SO! After all, you've got to make sure you've got an organization and team that is open to new ideas, and is looking at innovative ways of turning challenge into opportunity. In addition, you must make sure that you aren't an organization that is impacted by sudden new surprises – based on trends that are entirely and completely obvious.

Take the time to do this simple test. The objective is to determine if you and your team are in the right frame of mind for remarkably new and innovative things, or stuck in a rut, unable to respond and deal with the change that is swirling around you.

- Do your people laugh at new ideas?

- If someone identifies a problem are they shunned?

- Is innovation considered a privileged practice of a special group?

- The phrase, "you can't do that because we've always done it this way" is used for every new idea.

- Does anyone remember the last time anyone did anything really cool?

Does anyone remember the last time anyone did anything really cool?

- Do people think innovation is just about R&D, and forget that it's also about the way things are done, and about deliverables?

- Is the organization more focused on process than success?

- Are there lots of baby boomers about, and few people younger than 30, such that "change aversion" is the predominant culture?

- After any type of surprise – product, market, industry or organizational change – does everyone sit back and ask, "wow, where did that come from?"

If you are guilty of more than a couple of these indicators, then you've got an organization that has a significant change aversion in place – and likely, isn't actively monitoring the key trends that might impact you.

You need to sit back and think about how different the world is going to be five years from now. In a world of hyperchange, it will be very, very different – and that's why undertaking an honest self-assessment of your culture and attitude is critical.

Is your team in the right frame of mind to do what needs to be done to get there?

There's a big reason why I decided that my career expertise should be focused on linking future trends to the concept of innovation : it just seems so blindingly obvious that if we understand the trends that will impact us, we can innovate as appropriate to respond to them.

Given this simple premise, I continue to remain stunned by how many organizations have senior management in place that has no clue as to the future trends that might impact them, nor how to innovative around those trends. The result are sudden "surprises" that, with a little bit of common sense and foresight, could have been managed around quite adequately. It seems that common sense is not that common.

One of the best examples has to do with what we are now witnessing in the global energy sector: massive skills inflation (increase in the cost of skills) with resultant cost overruns, as the result of a shortage of workers, increasing skills specialization, and massive global competition for those skills.

The cost of the problem is staggering in some sectors: in the red-hot economy of Alberta, Canada, there are several massive projects that involve the extraction of usable oil from what are known as the "oil-sands." Some of the projects are faced with budgets that have ballooned from $4 billion to $8 billion; in some cases, even more.

Yet it isn't like there weren't signs that these issues were coming. The issue of the skills crisis has been on the table for quite some time; a scan through literature in the energy sector shows all kinds of examples in which industry experts were predicting significant challenges in the future.

Yet, if you have an organization that comes into a $4 billion cost over-run, isn't there some type of senior management negligence, particularly given that the potential problem was well known?

That's why an innovation audit – although it might sound trivial – is critical, because it will help to demonstrate whether your organization is awake or asleep; aware of the obvious trends that are coming, or simply blind to them.

Seeking Support

Communicate!

INNOVATION ALWAYS involves change, and dealing with that change is probably one of the biggest barriers to successful innovation. Introduce any new idea and you are bound to get the typical reactions, ranging from instant negativity, downright opposition and in some cases, misinterpretation of the reasons for the change.

The most important thing you need to do is:

Communicate, communicate, communicate –
and then communicate some more!

Time and again, change initiatives have failed as a result of a simple lack of communication. Poor communication leads to all the classic signs of an ineffectual change process, ranging from suspicion, fear, confusion to distrust and rumor.

All too often, those trying to effect the change have simply rammed the change through, without effectively explaining the reasons underlying the need for change, the implications of the change, and the benefits that will come from the change.

You will do much better if you spend as much time in creating a "change communication plan" as you do in structuring the details of the change itself. That will help to ensure that your message doesn't go off the rails. You have to continually ask yourself, 'am I failing to do this with my staff? with my customers? with my business partners?'

Anticipate objections

Many people seem to be driven by a rather simple outlook on life: whenever confronted with something new, they quickly respond with "we can't change things, we've always done it that way."

If you get right down to it, such a statement masks the reality of the fact that they just don't like change, don't want to have to deal with change, and certainly won't accept change! They've been born with a change anti-virus that immediately rejects any invasion of their comfort zone by any type of new initiative. The result – you will be guaranteed a great deal of grief as you attempt to move forward.

Your change strategy needs to take into account the fact that you will have strong objections to your plans, regardless of how small or large the nature of the change might be. You are going to have to be prepared to move your staff out of their comfort zone and into their 'discomfort zone.' Take the time to understand the potential objections, and then document and communicate how those concerns are likely unfounded. You might find that by doing this changes become a little easier to implement.

Make the presumption that you will need to educate the uninformed

For most people, their daily life is based upon routine – they come into work each day, and do the same thing that they did the day before – day in and day out. They live such a life of routine that they don't ever spend time thinking about trends, the future, and how their world will evolve around them – call them 'change-blind'. Because of their complete lack of any sort of change-radar, they expect that everything will always stay the same.

This is why you must work so hard to communicate the reasons for change. Most will not be in possession of the most basic facts related to the change. You will need to spend a lot of time not only communicating with them, but educating them.

Many people seem to be driven by a rather simple outlook on life: whenever confronted with something new, they quickly respond with "we can't change things, we've always done it that way."

Education implies information – and you can't have too little information.

Presume that you are dealing with indecision
Most people would prefer if they lived in a world in which there was never any change. When confronted with the need to change, they fall prey to that other common human instinct, an inability to make a decision!

Effecting successful change always requires a degree of support, yet that support will not be forthcoming if people are driven by the aggressive indecision that seems to be so characteristic of our times. Design your change initiative and communication plan in such a way that you are forcing people into making a decision.

This can be as simple as putting in place a firm deadline by which they must act, or by providing very clear choices that must be made.

Most people would prefer a world in which there was never any change. When confronted with the need to change, they fall prey to the common human instinct, an inability to make a decision.

**Plan for the fact that
people will misinterpret what you say**

Many change initiatives will run up the ingrained distrust that exists within the culture of many organizations. It's not surprising that many people will inherently distrust you – after all, there is no doubt that people have been battered by an extremely negative corporate environment through the last few years, particularly as corporate cost cutting has come to be the key or only change initiative of many an organization.

In such an environment, any type of change you propose might simply be viewed as a part of that cost-cutting agenda. You can counter this by being clear, succinct and concise. You also need to have your ears firmly plugged into the rumor mill, and must be prepared to act immediately on any misinformation that you might hear is being spread about.

Harness the passion of the supporters

Successful organizational change initiatives always involve the participation of those who are eager to see the change come about, or who have a stake-holding in the successful implementation of the change. You'd do well to get them involved as early as possible, since they will be very powerful allies.

Seek the involvement of the detractors

At the same time that you get the involvement of those who want to see change, you will find that you also need the support of those who are only marginally against the change.

You'll make things easier for yourself if you spend a bit of time to try to turn them into supporters. It might simply be a lack of information that has them sitting on the fence. Spending some time with them to understand their concerns and educating them on the benefits of the change will be well worth the effort. Segment your audience, address their concerns appropriately, and you might find that you are expanding your support base significantly.

Be positive but address the negative, don't try to sugarcoat

Hopefully, you are trying to effect some positive change, and can clearly outline a number of the benefits that will come once the change is in place. Even so, there is always a downside to any type of change, and you shouldn't hesitate in outlining that downside.

Make sure all of your communications address any negative issues without hesitation. Don't try to sugarcoat them – people will see through that, which can only help to fuel the negativity that can come about from a poorly managed change process.

Be honest, forthright and ethical.

Last but not least, ensure that your change initiative is based upon simple human decency and values.

There's a tremendous amount of workplace stress that revolves around change, and it can have a devastating impact on the ability of an organization to respond in the high-velocity economy.

While preparing for a change management workshop for staff of a government organization, I undertook some research into the issue of workplace stress. It didn't take long to come up with some facts that indicate just how bad things are: an article in the Daily Mail indicated that most of us spend about 2 1/2 hrs per day worrying about work and our career! Add that up over 40 years, and it comes to some three years and nine months of our life is spent fretting over work. It seems to be a bit worse for women: the same article went on to note that 65% of females admit to waking up in a cold sweat worrying about their jobs compared to 43% of men.

Then I found a book called "Manage Your Own Career: Reinvent Your Job, Reinvent Yourself," which carried this chilling statistic: "70% of people don't really enjoy what they are doing."

How can you possibly provide for a culture of innovation with such stunning negativity? By ensuring that people aren't kept in the dark about what is going on.

I was dealing with a major agricultural company that had a big "innovation drive" underway. They'd established a separate group of people, charging them with the responsibility of coming up with new ideas to deal with some of the unique market issues that are emerging, some of which would result in significant organizational and career change.

Yet they've made the classic change-management mistake: the mere fact that this group set off to do its work "secretly" turned on a corporate wide rumor mill.

The whole issue of change quickly spiraled out of control; people feared for loss of control or their jobs; there was widespread dislike and distrust of the message coming from management. There was suspicion and hostility, rumors and fearmongering.

If only the company had approached the issue of change with an open and clear mindset! The natural reaction of most people is to examine the threat of change, and not focus on the opportunity.

That's because when confronted with the unknown, our human nature instinct is one of "fight or flight."

Ready, Set, Done. • How to Innovate When Faster is the New Fast

Elasticity!

What makes for corporate agility?

THE CONCEPT of corporate agility is perhaps the most critical attribute that organizations need in the high velocity economy. Agility implies that we must innovate and adapt on a continuous basis due to rapidly changing circumstances.

How do we do that? By adopting several key guiding principles that form the basis for all corporate strategy and activities going forward.

- **Plan for short term longevity.** No one can presume that markets, products, customers and assumptions will remain static: everything is changing instantly. Business strategies and activities must increasingly become short term oriented while fulfilling a long term mission.

- **Presume lack of rigidity.** Many organizations undertake plans based on key assumptions. Agile organizations do so by presuming that those key assumptions are going to change regularly over time, and so build into their plans a degree of ongoing flexibility.

- **Design for flexibility.** In a world of constant change, products or services must be designed in such a way that they can be quickly redesigned without massive cost and effort. Think like Google: every product and service should be a beta, with the inherent foundation being one of flexibility for future change.

Think like Google: every product and service should be a beta, with the inherent foundation being one of flexibility for future change.

- **Build with extensibility (i.e. let others add onto your product).** Apple understood the potential for rapid change by building into the iPod architecture the fundamental capability for other companies to develop add-on products. Think the same way: tap into the world. Let the customer, supplier, partners and others innovate on your behalf!

- **Harness external creativity.** In a world in which knowledge is evolving at a furious pace, no one organization can do everything. Recognize your limits, and tap into the skills, insight and capabilities of those who can do things better.

- **Plan for supportability.** Customers today measure you by a bar that is raised extremely high – they expect you to deliver the same degree of high-quality that they get from the best companies on the planet. They expect instant support, rapid service, and constant innovation. If you don't provide this, they'll simply move on to an alternative.

- **Revisit with regularity.** Banish complacency. Focus on change. Continually revisit your plans, assumptions, models and strategies, because the world next week is going to be different than that of today.

The global "open source" movement is having a tremendous impact on every industry. It recognizes the reality that partnerships are a critical component of innovation, and that concepts such as "extensibility" – letting others add onto your product – are becoming extremely important.

Open Source originated in the software industry. Several like-minded individuals, with a passion for great software, harnessed the connectivity of the Internet to develop a wide variety of new software platforms, the best known of which is Linux.

Yet, the concept of "extensibility" and other ideas found here can apply to any type of product or service. One of the coolest devices I have in my home is a SqueezeBox, from the company Slimdevices.

Essentially, it's like an iPod except that it accesses the music found on my home network, and pumps that into our home stereo.

One of the founding principles of the company when they built the product – which consists of both hardware and software – was that it should be done in such a way that users could "add to the product." They did so by providing certain basic standards that software developers could use to enhance the product.

The result has been the emergence of a wide variety of "plug-ins" that add all kinds of new functionality to the product.

Just Do It!

How can you increase the importance of innovation?

HERE'S A quick list of nine things that smart, innovative organizations do to create an overall sense of innovation-purpose:

• **Heighten the Importance of Innovation.** A major company in the aerospace industry with several billions in revenue has eight senior VP's who are responsible for innovation. They don't just walk the talk – they do it. The message to the rest of the company? Innovation is critical – get involved. You can do the same thing by elevating specific responsibility for innovation to senior people throughout the organization, and then ensuring they deliver.

• **Create a Compelling Sense of Urgency.** With product lifecycles compressing, markets witnessing fierce competition, and entire skills and professions changing overnight, now is not the time for studies, committee meetings and reports. It's time for action. Simply do things now. Get it done. Analyze it later to figure out how to do it better next time.

• **Ignite Each Spark.** Innovative leaders know that everyone in the organization has some type of unique creativity and talent. They know how to find it, harness it, and use it to advantage.

• **Re-evaluate the Mission.** You might have been selling widgets five years ago, but the market doesn't want

Innovative organizations don't have management and staff that quiver at the thought of what might be coming next. Instead, they're alive from breathing the oxygen of opportunity.

widgets anymore. If the world has moved on, and you haven't, it is time to re-evaluate your purpose, goals and strategies. Rethink the fundamentals in light of changing circumstances. This is particularly important for associations in fast moving industries.

- **Build Up Experiential Capital.** Innovation comes from risk, and risk comes from experience. The most important asset today isn't found on your balance sheet – it is found in the accumulated wisdom from the many risks that you've taken. The more experiential capital you have, the more you'll succeed.

- **Shift from Threat to Opportunity.** Innovative organizations don't have management and staff that quiver at the thought of what might be coming next. Instead, they're alive from breathing the oxygen of opportunity.

- **Banish Complacency and Skepticism.** It's all too easy for an organization, bound by a history of inaction, to develop a defeatist culture. Innovative leaders turn this around by motivating everyone to realize that in an era of rapid change, anything is possible.

- **Innovation Osmosis.** If you don't have it, get it – that's a good rule of thumb for innovation culture. The same aerospace company lit a fuse in their innovation culture by buying up small, aggressive, young innovative companies in their industry. They then spent the time to carefully nurture their ideas and harness their creativity. You might be able to do the same type of thing with a merger with a similar, smaller and more innovative organization, or by seeking out some new, creative talent for your executive ranks.

- **Create Excitement.** There are too many surveys which indicate that the majority of people in most jobs are bored, unhappy, and ready to bolt. Not at innovative organizations! The opportunity for creativity, initiative and purpose results in a different attitude. Where might your organization be on a "corporate happiness index?" If it's low, then you don't have the right environment. Fix that problem – and fix it quick.

Innovation isn't a one time process – it's a cultural attitude in which everyone understands that they can play a role in running the business better, growing the business and transforming the business. It can be terribly difficult to get it started, which is why trying out some of these approaches can help you.

One of the most important things that you need to do with innovation is to make sure that it is routine, normal, and a constant part of everyday activity.

Most organizations fail to do so because of their inability to keep up with high-velocity change – and because they've made dealing with such change the responsibility of only a "privileged few."

You can easily spot such companies: they use suggestion boxes, instant brainstorming sessions, and other tricks that try to "create creativity." In this way, they practice "innovation elitism," and by doing so, they tend to trivialize innovation.

Yet, truly innovative organizations eschew such special effects. They've made innovation a core concept of every employee: it's in their job description! Their performance evaluation includes an assessment of the innovative ideas they have sponsored or have helped to implement. Bonus and reward systems have been adjusted so that innovation – both success and failure – are rewarded.

One client focused on this by highlighting the concept of "operational innovation," and the core goal of achieving "operational excellence."

The CIO, with the blessing of the CEO, set out several key innovation goals: let's remove any ongoing cost inefficiencies through the deployment of more sophisticated technology; let's increase customer loyalty and commitment through "fanatical customer service"; let's learn to be more collaborative, and thereby spot future market opportunities, by linking together our sales, marketing and other customer-oriented staff. Many other objectives such as these were highlighted and given priority.

Over a short time, staff came up with dozens and dozens of ideas to implement all kinds of small, operational, and "incremental" improvements.

In essence, they defined innovation as something that everyone could be involved in – and saw the results.

Degrees of Innovation

Focus on developing your Masters in Business Imagination

COMPLACENCY IN a time of rapid, disruptive change can be a death sentence – not only for organizations, but for the careers and skills of those who work there! It's time to abandon the thinking that has had you anchored firmly to the past – and to shift your focus to the future, with enthusiasm, motivation and imagination.

You can do this by abandoning any pretence that the skills of yesterday will be important tomorrow. Figuratively and literally, it is time to move beyond the thinking that has led us to a world of MBA's – Masters of Business Administration – and focus upon the critical skill that will take you into tomorrow.

The world doesn't need more administrators. It needs more MBI's – Masters of Business Imagination!

The criticality of change

Clearly you need different skills to take you into a future that is becoming far more complex, challenging and different by the minute. How can you keep operating the way you do – with the same culture, structure, rules and methodologies, when the rate of change that envelopes your organization is so dramatic and so darned fast?

We are in a time that demands a new agility and flexibility: and everyone must have the skill and insight to prepare for a future that is rushing at them faster than ever before.

Most people don't have such a capability; indeed, most people continue to stumble and meander their way into the future, without any firm grasp of how their skills, knowledge or industry is evolving. Without such insight, they have no idea of what they must do in order to thrive in an era of rapid change.

Worse yet, they focus on managing, rather than leading; administering, rather than inspiring; complying, rather than creating.

The result is that they continue to wake up each morning and think, "what happened to the world I knew?" Perhaps that is because their focus has been misdirected – they've become experts in "administration" at a time when what they really need is a lot more "imagination."

That's why progressive, future oriented leaders focus on developing their skills so that they can be honored with the title: Masters of Business Imagination!

The elements of an MBI

What are the core attributes and attitudes which individuals awarded with the degree possess? They have the ability to:

- **See things differently.** MBI's don't look at things like most people. They continuously challenge the assumptions that surround them, and use that as the formative fuel for their creativity. They know that the foundation of everything around them is shifting and twisting, and that it is in such movement that the future is being defined. They are willing to abandon any attitudes which might cause them to believe that everything is going to stay the same, with the result that they view the world through a different set of lenses. These lenses help them to generate new ideas, come up with imaginative solutions, and think creatively on a continuous basis.

- **Spur creativity in other people.** An MBI possesses a unique and critical skill: they can spur others around them to develop similar levels of imaginative thinking. In doing so, they can shake a team out of its administrative complacency, and motivate them into a mode in which they are able to rethink, redo and re-imagine, such that they lend much more value to the organization.

- **Focus on opportunity, not threat.** MBI's realize that in the absence of action, continuous disruptive change inevitably has negative consequences. With this being the foundation of their attitude, they have come to learn that their focus must continuously examine how to capitalize on change, in order to turn it into opportunity. They regularly scan for signs of disruptive threat, and rather than viewing it as something to be feared, ask themselves: "where is the potential here?"

- **Refuse to accept the status quo.** Ogden Nash, a great American poet known for his puns, once observed that "progress is great, but it has gone on far too long." Such stale thinking doesn't drive the passion of an MBI: instead, they look at the world around them, and constantly question how they might shake things up a bit, not simply to cause change, but in order to provide a climate in which the people within it can succeed and excel.

- **Bring ideas to life.** MBI's motivate people to excel by helping them to get over their own concern and worry about the future. They paint a picture of where the organization is going to go, and what it is going to take to get there. They provide forward oriented goals and objectives that are framed around the change that surrounds them, and use this as the fuel to spur their team on to achievement.

- **Learn and unlearn.** MBI's know that we live in a world in which learning has become their job, and just-in-time knowledge has become the foundation for future success. They don't make the dangerous assumption that what they know will carry them into tomorrow; they realize that the knowledge and skills that they will need to do their job in the future will require skills that are infinitely more complex. Rather than viewing this as a burden to be assumed, they view the opportunity to continuously learn something new with passion and enthusiasm.

- **Refuse to say the word "can't".** MBI's refuse to accept the limitations that might have been placed upon them. They know that barriers, perceived or otherwise, are simply temporary roadblocks that they can get around with fresh insight, imaginative analysis, and creative thinking.

- **Accept challenges with passion and enthusiasm.** Most studies continue to show that many people go to work each day with dread, fear and worry being their constant companion. Once they arrive, they fall into a monotonous routine of meetings, checklists and to-do's. Not MBI's: they approach each day as a new and exciting opportunity! They know that the world is wide open for them to use their way of thinking differently to shake things up, discover opportunity, and redefine parameters. Their passion as leaders is such that their enthusiasm becomes infectious to such a degree that they begin to steer the entire organization towards their fresh and exciting way of thinking.

- **Embrace change rather than shying away from it.** MBI's relish the idea of change, for it is the oxygen that fuels their fire every day.

- **Listen to people who are different than them.** An effective MBI knows that the attitudes that they possess can restrict them in their thinking, and that the knowledge they know is but a sliver of what there is to know. The result is that they are constantly on the prowl for new ideas and new ways of thinking, and they understand that often this can come from those who think differently than they do. Examine the make up of a team assembled by an MBI, and you will find a group of people who are all very different. That's because an MBI knows that a multitude of

Embrace change rather than shying away from it. MBI's relish the idea of change, for it is the oxygen that fuels their fire every day.

difference can bring far more creative thinking than the sameness from a group of clones.

- **Live for the opportunity to have ideas challenged and debated.** MBI's truly understand that their own bias, developed through years of experience, can often blind them to real opportunity. The result is that they are eager to have any idea assessed, analyzed and challenged. They encourage debate, knowing that it is through such a process that a simple idea can be turned into a "great idea."

- **Say "how can we make it work?" rather than "it won't work".** MBI's refuse to accept failure as an option, and instead, constantly work to ensure that goals are achieved. If they don't have what is needed to make it work, they look for creative and imaginative ways to get around that limitation.

An MBI is the most important degree program for any executive for the future. And although the degree isn't yet enshrined in any university curriculum – at least at the time this book was written – it's clearly the most important degree for the high velocity economy.

Closing

Closing

10 Great Words

IN CLOSING, what's the best thing that you can do to turn the future into opportunity, and accelerate your efforts for creativity and innovation?

Adopt ten simple words that will help to get you into the right frame of mind!

- **Observe.** Take the time on a regular basis to look for the key trends that will impact you, the industry you work within or the career you have established. Take the time to learn about the many automated knowledge discovery tools that exist on the Internet, a topic that is far beyond the scope of this book. Far too many organizations sit back after a dramatic change and asked "what happened?" Make sure that your organization is one that asks, "what's about to happen? And what should we do about it?"

- **Think.** Analyze your observations: spend more time learning from what you see happening around you. If you are like most organizations, you are responding to trends on a short term, piecemeal basis: you are reactive, rather than proactive. Step back, take a deep breath, and analyze what trends are telling you. From that, do what really needs to be done.

- **Change.** In a time of rapid change, you can't expect to get by with what has worked in the past – tin cans won't help you – you must be willing to do things differently. Abandon routine; adopt an open mind about the world around you. The world is changing at a furious pace whether you like it or not. Take a look at how you do everything – and decide to do things differently.

- **Dare.** Have you lost your ability to take risks? Maybe so – yet risk-taking is critical to innovation and change. Work with a few of the new ideas that you generate, and try them out. That's the only way that you will be comfortable with what comes next. (One friend suggested that you should 'fail fast, fail often, fail cheaply.')

- **Banish.** Get rid of the words and phrases that steer you into inaction and indecision: "We can't do that." "It won't work." "That's the dumbest thing I ever heard." These are the innovation killer phrases – watch for them, and don't permit them to be used.

- **Try.** How many of your people have lost their ability to adapt to changing circumstances because they've lost their confidence? Developing new skills and career capabilities is critical, given the rapid change occurring around us. Yet too many people have managed to convince themselves that they can't adapt; they can't change. Don't let that happen – it's one of the worst attitudes for going forward into the future.

- **Empower.** In a world of rapid change, you can't expect that rigidly defined rules will be the appropriate response to changing circumstances. A ticked off customer needs a solution right now from a front line customer service rep – not some type of follow-up from head office weeks later. A middle manager in a remote location needs the ability to make a decision and must commit to it today – they can't afford to wait for the wheels of head office bureaucracy to churn. Destroy the hierarchy, and re-encourage a culture in which people are given the mandate and the power to do what's right, at the right time, for the right reason.

- **Question.** Go forward with a different viewpoint by challenging assumptions and eliminating habit. If your approach to the future is based upon your past success, ask yourself whether that will really guarantee you similar results in the future. If you do certain things because "you've always done it that way," then now is an excellent time to start doing them differently.

- **Grow.** Stop focusing strictly on cutting costs – focus on service opportunity instead. Don't stand in fear of what you don't know – teach yourself something new. Don't question your ability to accomplish something great – grab the bull by the horns and see what you can do! The point is, in a world of rapid change, you must continually enhance your capabilities and opportunities through innovative thinking. Change your attitude now, and the rest will come easily.

- **Do.** Renew your sense of purpose, and restore your enthusiasm for the future by taking action. Too many organizations, and the people who work within them, are on autopilot. They go into work each day, and do the same things they did the day before, with the belief that everything today is the same as it was yesterday. It isn't.

Oh, and there's an 11th word: **Enjoy.** Through the years, I've come to learn that the groups that pursue innovation are those organizations and individuals who approach the future with a lot of passion. These are the folks who tend to wake up every day and think, "wow, I can't wait to get to work!".

Rapid times require bold change; action is critical. Confront your tin cans, and you've got the right leadership frame of mind to take you into the future.

Ready, Set, Done. • How to Innovate When Faster is the New Fast

Acknowledgments

Every book that I've worked on through the last decade has involved a unique set of challenges. This one was perhaps the most unique, in that the process of conceptualizing what the book was to become went through so many different twists and turns.

To my wife Christa, for once again playing the role of editor and "thought-shaper." The patience she displays in simply trying to comprehend what I am talking about – whether within the book or in the real world – is remarkable. And for working with me in the home office for 15 years, she deserves some sort of medal!

To my sons Willie and Thomas, for their fascinating comments on the gazillion potential book titles we came up with ("too boring," "too weird," "yuck!") as well as their ongoing encouragement to get the book done so I could spend more time with them! This is what home offices are meant to be!

Rob Mustard once again pitched in with a review of the first draft of the book, and offered up many wonderful suggestions. Mark Jeftovic of easyDNS also deserves thanks for sharing his comments on the first draft. He also played a key role by introducing me to his bandmate Phil Emery, of Focused Creative Communications, who ended up designing the cover and internal layout for the book! Small connections make for great partnerships!

Last but not least, much of my insight has come from my client base. I am always grateful for the time I have spent with many fascinating people, learning from them, gaining insight into their creativity, and understanding what they are doing to succeed in the high velocity economy.

About the Author

Jim Carroll is a strategic thinker and "thought leader," with deep insight into trends, the future, creativity, and innovation. Named by *Business Week* as one of four leading sources for insight on innovation and creativity, he was also a featured expert on the prime time CNBC series, "*The Business of Innovation*," hosted by Maria Bartiromo.

Jim has a client base that includes the likes of Walt Disney Corporation, Nestle, Motorola, the American Society for Quality, Caterpillar, Verizon, the BBC, Blue Cross Blue Shield, the Property and Casualty Insurance Association of America, and the Swiss Innovation Forum.

An author, columnist, media commentator, and consultant, he has spent the last two decades providing direct, independent guidance to a huge, diverse client base. Prior to this he spent 12 years with the world's largest professional services firm.

Jim has researched key innovation success factors for dozens of industries, associations, professions, and companies, and provides high energy keynotes for audiences of 3,000, or intimate, detailed customized strategic planning workshops for CEO / board / senior management meetings.

He has provided his insight for the life sciences, health care, insurance, automotive, manufacturing, agriculture, technology, education, government, consumer products, retail, banking, and countless other industry sectors.

Made in the USA
Lexington, KY
26 October 2011